This festschrift volume is dedicated to William Liller, a noted observational astronomer who is now retired from a Harvard University named professorship. Nineteen of Liller's colleagues and former students (as well as Liller himself) give insightful reviews of the current state of a broad range of frontier astrophysical areas from the solar system to the limits of the observable universe. The reviews are of special interest to professional astronomers and students of astronomy, and are also accessible to the interested nonspecialist reader. The articles are well suited for providing graduate students in astronomy with an introduction to these topics. The authors describe their personal involvement in important research advances and convey a sense of the excitement of scientific discovery. The reader is given an overview of the advanced techniques used by today's observational astronomers for probing the mysteries of the cosmos, and of the theoretical interpretation of the observational findings.

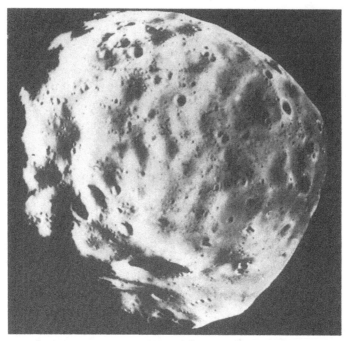

Phobos, a satellite of Mars approximately 29 km long, illustrates the expected appearance of asteroids. Phobos may have originated in the asteroid belt and then later have been captured into orbit about Mars.

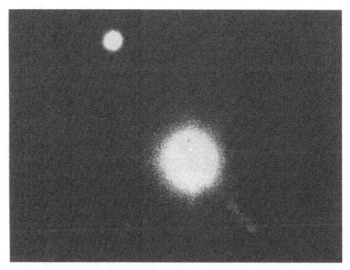

Quasar 3C 273 has a starlike appearance in this optical image, with a luminous jet to one side. Quasars, the most distant and luminous objects known in the universe, are believed to be the extraordinarily active nuclei of galaxies.

Asteroids to Quasars

A Symposium Honouring William Liller

Edited by

PHYLLIS M. LUGGER

Associate Professor, Department of Astronomy,
Indiana University

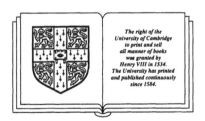

CAMBRIDGE UNIVERSITY PRESS
Cambridge
New York   Port Chester
Melbourne   Sydney

PUBLISHED BY THE PRESS SYNDICATE OF THE UNIVERSITY OF CAMBRIDGE
The Pitt Building, Trumpington Street, Cambridge, United Kingdom

CAMBRIDGE UNIVERSITY PRESS
The Edinburgh Building, Cambridge CB2 2RU, UK
40 West 20th Street, New York NY 10011–4211, USA
477 Williamstown Road, Port Melbourne, VIC 3207, Australia
Ruiz de Alarcón 13, 28014 Madrid, Spain
Dock House, The Waterfront, Cape Town 8001, South Africa

http://www.cambridge.org

First published 1991
First paperback edition 2004

A catalogue record for this book is available from the British Library

ISBN 0 521 35231 2 hardback
ISBN 0 521 61704 9 paperback

CONTENTS

History, Lore, and Archaeoastronomy

Indexes

ASTEROID DEDICATION

MINOR PLANET 3222 LILLER

Discovered by Edward Bowell on 1983 July 10 at the Anderson Mesa Station of the Lowell Observatory and presented to William Liller on the occasion of the Symposium: "Asteroids to Quasars", held in honor of his sixtieth birthday.

Citation:

Named in honor of William Liller, formerly the Robert Wheeler Willson Professor of Applied Astronomy at Harvard University, on the occasion of his sixtieth birthday. A premier observer, he has made substantial contributions through observations of a broad range of astronomical objects and phenomena: planetary nebulae, minor planets, comets, novae, variable stars, globular clusters, x-ray sources, quasars, solar eclipses and stellar occultations. Now living in Chile, he has in recent years participated in the PROBLICOM survey and has discovered several novae. During

(*Photo by* Marjorie Nichols)

the recent return of Halley's Comet, he was a crucial member of the IHW Island Network. He has been a leader in astronomical education and has been an important supporter of amateur astronomy. His enthusiastic encouragement has been greatly appreciated by his colleagues and students.

(*Photo by* Marjorie Nichols)

William Liller: Biographical Sketch

As a dedicated researcher and teacher, Bill Liller has made major contributions to a wide spectrum of astronomical fields including studies of planetary nebulae, comets, asteroids, magnetic activity in cool stars, and optical identifications of X-ray sources — to cite only a few. Following are some highlights of his accomplishments as kindly provided by Yervant Terzian, Freeman Miller, Jim Elliot, Andrea Dupree, Jay Pasachoff, Josh Grindlay, and Andrew Fraknoi.

Bill's Ph.D. dissertation was carried out with Lawrence Aller at the University of Michigan in the early 1950's on the topic of planetary nebulae. Bill built a photoelectric photometer to carry out this work, demonstrating his notable facility for developing and working with instrumentation. He made important contributions to the photometric observation of the central stars of planetary nebulae, to the study of the spectral changes of these objects, and to the measurement of their linear optical polarization. He was also a pioneer in the study of the angular expansion of nebulae in order to derive accurate distances.

While at Michigan in the late 1950's, Bill began collaborations with Freeman Miller on the study of comets. Bill carried out studies of neutral molecules and dust grains in comet comae and tails, using a photoelectric spectrum scanner that he had built, which was mounted on the Michigan Curtis Schmidt telescope. Bill continued his comet collaborations with Freeman Miller during the 1985-86 passage of Halley's Comet. They observed the comet with the venerable Curtis Schmidt, which was by this time located on Cerro Tololo in Chile. Bill also made extensive observations of the comet as part of the NASA International Halley Watch using a Celestron Schmidt which he set up on Easter Island. The first comet discovery of 1988 was made by Bill with the same Celestron, then mounted in the yard of his home in Reñaca, Chile. Comet 1988a was 13th magnitude at discovery, but reached naked eye brightness as it moved north.

Bill's talents in instrument development made possible the early successes in stellar occultation observations carried out with Joe Veverka and Jim Elliot. Bill's portable multi-channel photometer used for the occultation of β Scorpii by Jupiter in 1971 facilitated the acquisition of a high quality data set. The detailed analysis inspired by these data led to significant developments in the use of stellar occultations for studying planetary atmospheres, planetary ring systems, and asteroids.

In the area of cool stars, Bill initiated a line of research which was the harbinger of much contemporary work in magnetic activity of late type stars. While on sabbatical from Harvard at Cambridge University in 1964–65, Bill observed the emission cores of the Ca II K line in several stars, finding these to vary. This led to a program with Sallie Baliunas, Lee Hartmann, and Andrea Dupree to observe several stars with the chromophotometer on the Mt. Wilson 100 inch telescope, that resulted in the discovery of short time-scale periodicities in several luminous stars. This work was followed by the ongoing large scale Smithsonian-Mt. Wilson program to monitor several hundred stars.

Bill has always been interested in seizing the opportunity to do new and exciting research. For example, he branched into solar physics when a chance arose to observe the nighttime sky for an exceptionally long period during the daytime as a result of the total solar eclipse of 1973 in Africa. Bill's teams went to Mauritania and Kenya as part of the U.S. national expedition.

In the early 1970's Bill Liller led the effort at the Harvard-Smithsonian Center for Astrophysics to optically identify X-ray sources recently discovered by the *Uhuru* satellite. As noted by several contributors to this volume, he served as mentor to a generation of X-ray astronomers. Bill's discovery of the optical counterpart to Her X-1, in 1972, paved the way for the detailed understanding of high luminosity galactic X-ray sources as compact binary systems in which a neutron star accretes gas that spills over from its normal companion. In 1977 using the newly available IV-N infrared sensitive photographic emulsions, Bill discovered a highly obscured globular cluster (now known as "Liller 1") containing the unique "rapid burster" X-ray source. Bill also contributed to the development of extragalactic X-ray astronomy, by identifying ultraviolet bright optical counterparts to faint Einstein X-ray sources; these objects were shown to be quasars.

Reflecting his continuing diverse interests, Bill's recent research activities from his home base in Chile include successful nova searches, determinations of the ages of globular clusters, and studies of the archaeoastronomy of Easter Island.

In addition to his impressive list of research accomplishments, Bill has also made significant contributions in the area of astronomical education. These include his popular undergraduate introductory astronomy courses at Harvard, his patient and good-humored counseling of several generations of students, his encouragement of amateur astronomers, his co-authorship with Ben Mayer of the *Cambridge Astronomy Guide*, and his articles for *Sky and Telescope* and *Mercury*.

In an era of increasing specialization, Bill has been the consummate generalist, making important contributions in a broad range of areas. His evident delight in everything that he does is an inspiration to us all.

Phyllis Lugger
Indiana University

AUTOBIOGRAPHICAL MEANDERINGS

W. Liller
Instituto Isaac Newton, Ministerio de Educacion
Casilla 8-9, Correo 9, Santiago, Chile

Hardly born in humble surroundings (my father was in the advertising business in Philadelphia), I first became obsessed with astronomy in late August 1932. My intellectual uncle had driven all the way from Keyser, West Virginia in an A-model Ford to see the total solar eclipse that was to pass (on the 31st) well north of suburban New York City where we were then living. Two things impressed me mightily: the fact that Uncle J.W. travelled so far, and the sight, a few days later, of the partial eclipse.

My parents encouraged my early astronomical obsession, rewarding me (at age 12) with a 2-1/2 inch refractor ("Astron-O-Set") after I somehow struggled through Rachmaninov's Prelude in C-sharp Minor without a mistake.

My first publication -- or rather citation -- came the next year when I reported seeing several dozen Quadrantid meteors from Atlanta, Georgia where we had moved when I was 11. (The head of the American Meteor Society, C.P. Olivier, obviously miffed that I had mis-spelled his name, refers to me in Popular Astronomy as Billy Lillier.)

Believing the Atlanta public high schools to be something less than rigorous, my parents then sent me away to a Pennsylvania boarding school that I much disliked partly because the teaching (especially math and science) was, I felt, inferior to that of my Atlanta public high school (Drs. Avrett and Lightman: take note!). Moreover, no one there had the slightest interest in astronomy. More encouragement was given to my baseball prowess, to which, I must admit, I did not object.

One day my best friend asked me where I was going to college; I shrugged. He said he was applying to Harvard; O.K., I would too. Luckily, we both were admitted; neither of us had applied elsewhere. (My friend is now a Professor of History at the University of Michigan.)

We were freshmen together in Adams House beginning in June 1944 (there was a war on), and I divided my time between variable star observing for the AAVSO, baseball (or soccer), girls, and when there was time, course work. A year later I was drafted and sworn into the U.S.Navy as a Radio Technician. Eleven months and five days later I was sworn out of the U.S.Navy, having learned absolutely zero about radios. However, I had possibly become the fastest man in the East on that venerable but now obsolete device called a slide rule. And I also had developed an intense dislike for anything military.

Back to Adams House I went and settled into the Astronomy Department where I had for advisers in each of my last three years Professors Bok,

Menzel, and Whipple, all of whom strongly influenced me in one direction or another. So did a number of my fellow (mostly senior) students: Hoag, King, two Wyatts, Harlan Smith, McCrosky, Arp, Newkirk, to name a few. And the Director, Harlow Shapley, influenced us all.

In the summer of 1948 Dr. Whipple asked me to help put together an expedition to New Mexico to photograph meteors, and I took the following semester off living with Harlan Smith and Dick McCrosky, also of the Meteor Expedition, in a basement on the wrong side of Beacon Hill. It was during these times that Harlan and Dick developed their widely known "cherry pit pie". We also formed a singing group that acheived local fame (or infame) in Observatory circles and was known simply as Smith, McCrosky and Liller.

Had she come along a year or two earlier, I might have switched my field of concentration to music, because Elizabeth Menzel (now Mrs. Bernard Davis) gave me the confidence (and much motivation) to enroll in the beginning theory course in my senior year, and the instructor, now Prof. Richard French at Yale, was one of the best and most influential of all my teachers.

Frances Wright, very much the musical director at the Observatory, organized a full symphony orchestra (the Observatory Philharmonic Orchestra), except for a bass violin, which I eventually constructed out of 2-by-4s and plywood from the Observatory carpenter's shop and the Amateur Telescope Makers' shop which was then in the basement of Building "A". The first OPO concert, myself on bass, was performed in Shapley's living room ("Formal Dress Optional"; Babbie and Fred Whipple -- and a few others -- were resplendent in evening gown and tuxedo). This gala event is described elsewhere, but I will mention here that a string quartet composed of Cecilia Gaposchkin, David Layzer, Joyce Marrison, and Elizabeth Menzel performed "Isostasy", a work composed by yours truly. (Part of it had been a homework assignment in the music course, as Prof. French immediately recognized; he, too, was present.) As I understand it, Art Hoag's recording of the concert has not yet been submitted to the Library of Congress.

The hot field of astronomy at the time was stellar atmospheres, and Leo Goldberg and Lawrence Aller, authors of the popular book "Atoms, Stars, and Nebulae", were both at the University of Michigan. Although I was sorely tempted to remain at Harvard for graduate work (I had done my undergraduate thesis --"Radio Detection of Meteors at 3.5 Mc/s" -- with "Uncle Fred" Whipple), the lure of Ann Arbor won out. I arrived there in the fall of 1949.

I had had a very good time at Harvard, so good that my grades were less than sterling, and my new academic year's resolution in Ann Arbor was to forbid myself the pleasures of concerts, movies, girl friends, etc. for the first year (except for an occasional trip back to Cambridge). This strict regimen improved grades to the point where I could relax somewhat my monkish ways in the second year. My office window (shared with three other grad students) afforded an excelent view of the new girl's dormitory next door, and one third-year student directly across from us, Lorraine Dundas, was particularly meddlesome in a most attractive way. We were married in April of 1951.

Prof. Freeman D. Miller, noting that I had a certain talent building instruments (I had built a clavichord in my spare time 1948-1952), offered departmental support if I would design and build a photoelectric photometer, something that was conspicuously absent from the Observatory's inventory. And Prof. Aller suggested photometry of planetary nebulae as a first application of the new instrument. All of a sudden I had a thesis project, soon a working photometer, and observations were underway.

Meanwhile, back in Cambridge, Uncle Fred was searching for someone to take over as "superintendent" of the Harvard Meteor Expedition, now operating smoothly, thanks mainly to Mssrs. Smith and McCrosky. As soon as I had accumulated all the necessary nebular (and central star) observations with the new photometer, Lorraine and I headed (June 1952) to Las Cruces, NM where I divided time between superintending the station, writing my dissertation, and doing a little amateur archaeology.

Meanwhile, back in Ann Arbor, "Uncle Leo" Goldberg was searching for a junior faculty member, and offered me an instructorship, beginning in September 1953. Back we went, and there I stayed put for seven years, working with Aller (on planetary nebulae), Miller (on comets), and Goldberg (on the sun), while steadily moving up through the academic ranks. My instrumental interests continued, and the design and construction of a photoelectric spectrum scanner led to several publications and my first PhD student, Robert Bless.

As it did to so many astronomers, Sputnik drastically changed my life. Leo Goldberg asked me to work with him on a solar satellite experiment, and Lorraine, understandably, found she could no longer put up with my astronomical obsession.

Once again back in Cambridge, Donald Menzel, now Director at Harvard, had negotiated two faculty positions and he offered them to Leo and me. At the time (1960) Michigan institutions were suffering because car sales were down, and so back to Cambridge we went. (Leo was a Harvard product, too.) With me was my bright and pretty new wife, Martha Hazen, a recent recipient of a PhD from the University of Michigan.

No sooner had I moved into my new office when Donald Menzel asked me to be departmental chairman succeeding Cecilia Gaposchkin who was just finishing her term of office. To follow her -- replacing her was impossible, of course -- was an honor I could not resist; I had long been an admirer of the great "CPG".

My title was the Robert Wheeler Willson Professor of Applied Astronomy (see the article by Barbara Welther in this volume), and when I expressed concern about one of the clauses in the description of the duties of the occupant of this chair, specifically, that I should teach navigation (Frances Wright was very much in charge of that department), Donald suggested that I should say a few words each year about, for example, navigating a space probe to Mars or Venus.

The next several years were exciting and immensely busy. A U.S. Senator, John F. Kennedy was the Chairman of the Astronomy Department Visiting Committee (he resigned when he took higher office in 1961);

Richard Nixon was chairman of the SAO Visiting Committee (he had to step
down when John Kennedy became president); and many of my faculty
colleagues were young and vigorous, people like Dave Layzer, Ed Lilley,
Mort Roberts, Chuck Whitney, and a few years later Alex Dalgarno, Carl
Sagan, and Owen Gingerich (and again I cannot name all).

In 1964-65, thanks to a Guggenheim Fellowship, Martha and I took a year's
leave and settled into a 500-year-old house not far from the present
offices of the Cambridge University Press. That period was highly
stimulating and most pleasureable in a large part thanks to my Cambridge
University colleagues that included R.O.Redman, Roger Griffin, David
Dewhurst, Rodney Willstrop, and Noah Argue, plus visitors to "The
Observat'ries" like Burbidge, Burbidge, Fowler and Hoyle. Spare time was
again spent doing amateur archaeology, and my first paper in that field
was published (on Roman Roads in Britain). It was an unforgetable year in
a beautiful and invigorating environment.

Return to Harvard and teaching resulted in several more first-rate PhD
students (including Stephen Strom and Richard Berendsen). A memorable
part of this era was the graduate student suppers at our home in Belmont.
Entertainment always seemed spontaneous; at one party Carl Sagan boasted
that he would take on all comers at ping pong (at least two of us beat
him); another time, a horde of semi-tame raccoons came to the house
hoping to join in the cook-out, and brought a friend -- a skunk.

Subsequent retirement from the chairmanship led to a new lively responsi-
bility: I became (June 1968) Master of my undergraduate residence, Adams
House (the first ever to become Master of his own House). The first year
was difficult, to put it mildly: the storm, caused by student unrest over
the worsening Vietnam situation, reached its full fury the following
April.

But the uneasy calm that lasted for the next several years had its
rewards. Making up the list of Fellows and Associates of the Adams House
staff were faculty members from all departments including some of
Harvard's most illustrious scholars who frequently came to the House to
discuss their work or House problems and to participate in seminars,
plays, concerts, and to take tea and play House sports. (I was catcher
the year we almost won the baseball championship.) In 1970 the Master of
one of the Radcliffe women's Houses and I proposed to the Harvard
administration that we make an experimental swap of 50 students, thereby
making the two Houses "cohabitational". The experiment was a resounding
success and within a year, all the houses followed our lead and became --
and still are -- co-educational.

X-ray astronomy blossomed near the end of my 5-year term as Master, and
the '70s were devoted to working closely with Riccardo Giacconi and his
colleagues on the identification and understanding of the new sources.
It was yet another tremendously exciting period of my life, made even
more so by the collaboration with new PhD students such as Christine
Jones, Bill Forman, and Phyllis Lugger, while others like Debbie
Elmegreen and Bruce Carney kept me honest and happily involved in other
fields.

The frequent trips to observe at Cerro Tololo caused the next significant alteration in my life. I had begun to collaborate with the well-known Chilean astronomer Gonzalo Alcaino, and noting that one could retire from Harvard at the age of 55 and retain pension, health benefits, and use of the Faculty Club facilities, I decided in 1981 to pull up stakes and head south. Separation and subsequent divorce ultimately led to marriage (April 1985) with a lovely, vivacious Chilean widow Matilde Pickhardt, and we now live in a house just north of Viña del Mar with views of both the Pacific and the Andes.

Retirement provided the time to do some serious composing, and friend-ships with several of the finest organists in Chile have inspired me to write primarily for that magnificent instrument. I have been fortunate enough to have some of my efforts performed both here in Chile and abroad (USA, Belgium, and by now possibly Germany.)

As a "born-again" amateur astronomer, I began a routine patrol of the fabulously rich Southern Milky Way with an ordinary Nikon, and with a blinking technique developed by friend, co-author, and obsessed amateur astronomer, Ben Mayer, I have been lucky enough to catch six novae. (I missed a possible independent discovery of Supernova 1987A because, alas, it was cloudy *that* night in Viña del Mar.)

Gonzalo Alcaino heads a small but impressively productive research organization, the *Instituto Isaac Newton*, supported by the *Ministerio de Educacion de Chile*. Through his kindness, I was appointed Associate Director, and our scientific collaboration has continued to be an exceedingly pleasant and productive one. Our joint work is described by him elsewhere in this volume.

My past cometary collaborations with Freeman Miller led to a joint project on the tail structure of Halley's Comet and involvement in the International Halley Watch. During the comet's brightest phase, while Freeman observed at Tololo, I took photographs from Easter Island using an 8-inch Schmidt Camera provided by NASA. The runs were very successful for us both owing in part to good weather, and our photographs and scientific results have received wide attention.

On Easter Island, a new turn of interest occurred. On the island, I was approached by two distinguished archaeologists, Don Sergio Rapu Haoa, the Island Governor, and Dr. Georgia Lee from the University of California and asked to measure the orientations of various structures thought, perhaps, to be solstitially oriented. These investigations have opened up a whole new fascinating field for me and one which I am very much involved in now: the archaeoastronomy of Easter Island -- and other Polynesian islands as well. It is, I hardly need to add, exceedingly pleasant work.

Looking back: The above paragraphs should make it obvious to any reader that I have been extremely happy throughout my three score years. The fascination of astronomy, the stimulation of students (including a few thousand undergraduates), and the good fortune to have inspiring and enjoyable professors and colleagues have made my life intensely pleasureable. While I regret that, until now, I have not been able to find the marital harmony that I so much envied in other couples, I have the greatest admiration for the three remarkable women who have shared my life and given me their love -- and received mine. Finally, Tamara, John, and Hilary, my much beloved children, have made me immensely proud, and have given me infinite delight and many unforgettable memories.

Forty years ago, a business associate of my father told me quite bluntly that I was "a goddamned fool to go into astronomy". He could not have been more wrong.

(*Photo by* Tamara Liller)

PREFACE

On June 6, 1987 — a gloriously sunny spring day — fifty of Bill Liller's colleagues, students, family, and friends convened at the Harvard-Smithsonian Center for Astrophysics for a day-long symposium *Asteroids to Quasars,* held in honor of his sixtieth birthday. The papers presented reflected Bill's broad research interests, from the solar system to the limits of the observable universe. This festschrift underscores Bill's wide ranging impact on astronomy over the past four decades.

I would like to take this opportunity to thank those who helped make the Liller Symposium such a happy occasion. The symposium was made possible by the dedicated efforts of the local organizing committee: Christine Jones, Bill Forman, Josh Grindlay, and Jim Elliot. I am grateful to Christine and Bill for taking care of a myriad of organizational details. They masterfully orchestrated a galaxy of culinary delights — from the delectable coffee breaks through the delicious Chinese banquet lunch at Change Sho to the outstanding dinner at Adams House prepared by Cuisine Chez Vous. I would like to thank Josh Grindlay for providing financial support from the Harvard Astronomy Department and for his experienced assistance with symposium organization. I greatly appreciate Jim Elliot's many contributions. He arranged the naming of Minor Planet 3222 for Bill, wrote the citation, and designed the attractive certificate. His after dinner anecdotes and slide presentation about global observatory travels with Bill were delightful. Special thanks are due to Ted Bowell for naming his minor planet for Bill.

I wish to thank Yervant Terzian for suggesting the kaleidoscope gift to entertain Bill on cloudy nights and for selecting such a beautiful one for him. I also appreciate his masterful delivery of the paper by Lawrence Aller who was unable to attend due to a family illness. I am grateful to Owen Gingerich for presenting his very entertaining after dinner talk on prize winning spectral sequence mnemonics from Natural Sciences 9 (the contest had been originated by Bill). Owen had been in New York City for his son's wedding and drove back to Cambridge especially for the symposium banquet.

I would like to express my appreciation to Irwin Shapiro for making the facilities of the Harvard-Smithsonian Center for Astrophysics available to the symposium. I am grateful to his assistant, Judy Glas, for her help in organization. I would like to thank Robert Kiely, Master of Adams House, for providing use of the house facilities for the banquet. His assistant, Victoria Macy, was a great help in the preparations for the dinner. I wish to thank John Huchra for preparing the campus map that helped everyone find Adams House.

I am grateful to Steve and Karen Strom for suggesting the Boston University Photographic Resource Center which in turn recommended Marjorie Nichols, the symposium photographer. An album of her superb candid photos was presented to Bill as a remembrance of the day. I would like to thank Tom Bertolacinii for his skillful printing of the candids by Marjorie used in this festschrift. Thanks are due to Dennis di Cicco for his excellent group photograph of the conference participants.

I would like to thank my husband, Haldan Cohn, for his indefatigable assistance including his fine projection work and coffee preparation. I am grateful to my mother, Dorothy Lugger, for looking after our daughter, Alison, during the long day of the symposium. Thanks go to Alison for being a good sport. I am most appreciative of the many contributions made to my education by my mother and by my late father, Albert Lugger. I am grateful to my father for teaching me the importance of a kind heart.

Many thanks are due to Paula Jentgens and Brenda Records, the Indiana University Astronomy Department secretaries, for their excellent work in preparing the symposium mailings and for their help with the other symposium preparations. I would like to thank Dick Durisen for the financial assistance of the Indiana University Astronomy Department.

I am grateful to the chairs of the sessions, Wes Traub, Gene Avrett, Lanie Dickel, and Dan Harris for directing the lively interchanges. I wish to thank those participants whose contributions appear in this volume, as well as Sallie Baliunas, Steve Strom, and Cos Papaliolios, for stimulating presentations. I am grateful to all of the participants for making the day so memorable. I would like to thank all those who sent telegrams, letters, and photos to Bill for the occasion; many of these were shared at the banquet and all were warmly received. Thanks are due to Matty and John Liller for their cheerful presence.

Once the meeting was over, there was of course the "minor" task of assembling the proceedings volume. Thanks are due to Haldan for his continued assistance in this long work. George Turner's introduction to the wonders of TeX are appreciated. I would like to express my thanks to the contributors for their careful preparation of manuscripts and their agreeableness in making revisions. I am especially grateful to my physicians, John Kincaid, John Hayes, Valerie Purvin, Bernard Lown, and Howard Weiner for their thoughtful assistance during this time.

And finally, I would like to thank Bill for his countless insights and constant encouragement.

Phyllis Lugger
Indiana University

LILLER SYMPOSIUM
ASTEROIDS TO QUASARS
June 6, 1987
Harvard-Smithsonian Center for Astrophysics

PARTICIPANTS

Gonzalo Alcaino
Instituto Isaac Newton
Ministerio de Educación de Chile
Santiago, Chile

Lawrence Aller
Department of Astronomy
University of California
Los Angeles, California USA

Gene Avrett
Harvard-Smithsonian Center for Astrophysics
Cambridge, Massachusetts USA

Sallie Baliunas
Harvard-Smithsonian Center for Astrophysics
Cambridge, Massachusetts USA

Lola Chaisson
Harvard, Massachusetts USA

Haldan Cohn
Department of Astronomy
Indiana University
Bloomington, Indiana USA

Elizabeth Davis
Belmont, Massachusetts USA

Dennis di Cicco
Sky Publishing Corporation
Cambridge, Massachusetts USA

Lanie Dickel
Department of Astronomy
University of Illinois
Urbana, Illinois USA

Elaine Elliot
Wellesley, Massachusetts USA

Jim Elliot
Department of Earth, Atmospheric, and Planetary Sciences
and Department of Physics
Massachusetts Institute of Technology
Cambridge, Massachusetts USA

Debbie Elmegreen
IBM Watson Research Center, Yorktown Heights
and Vassar College Observatory, Poughkeepsie
New York, USA

Vladimir Escalante
Harvard-Smithsonian Center for Astrophysics
Cambridge, Massachusetts USA

Bill Forman
Harvard-Smithsonian Center for Astrophysics
Cambridge, Massachusetts USA

Owen Gingerich
Harvard-Smithsonian Center for Astrophysics
Cambridge, Massachusetts USA

Josh Grindlay
Harvard-Smithsonian Center for Astrophysics
Cambridge, Massachusetts USA

Sandy Grindlay	Lincoln, Massachusetts USA
Barbara Harris	Harvard University Cambridge, Massachusetts USA
Dan Harris	Harvard-Smithsonian Center for Astrophysics Cambridge, Massachusetts USA
John Huchra	Harvard-Smithsonian Center for Astrophysics Cambridge, Massachusetts USA
Christine Jones	Harvard-Smithsonian Center for Astrophysics Cambridge, Massachusetts USA
Jun Jugaku	Tokyo Astronomical Observatory Tokyo, Japan
Jacqueline Kloss	Cambridge, Massachusetts USA
Larry Liebovitch	College of Physicians and Surgeons Columbia University New York, New York USA
Bill Liller	Instituto Isaac Newton Ministerio de Educación Santiago, Chile
John Liller	Worcester Academy Worcester, Massachusetts USA
Matty Liller	Reñaca, Chile
Phyllis Lugger	Department of Astronomy Indiana University Bloomington, Indiana USA
Janet Mattei	American Association of Variable Star Observers Cambridge, Massachusetts USA
Mike Mattei	American Association of Variable Star Observers Cambridge, Massachusetts USA
Karen Meech	Institute for Astronomy University of Hawaii Honolulu, Hawaii USA
Florence Menzel	Cambridge, Massachusetts USA
Nancy Murphy	Harvard-Smithsonian Center for Astrophysics Cambridge, Massachusetts USA
Alice Papaliolios	Carlisle, Massachusetts USA
Costas Papaliolios	Department of Physics Harvard University Cambridge, Massachusetts USA

Jay Pasachoff	Hopkins Observatory Williams College Williamstown, Massachusetts USA
Naomi Pasachoff	Williamstown, Massachusetts USA
Leif Robinson	Sky Publishing Corporation Cambridge, Massachusetts USA
Rudy Schild	Harvard-Smithsonian Center for Astrophysics Cambridge, Massachusetts USA
Jerome Shao	Harvard-Smithsonian Center for Astrophysics Cambridge, Massachusetts USA
Karen Strom	Department of Physics and Astronomy University of Massachusetts Amherst, Massachusetts USA
Steve Strom	Department of Physics and Astronomy University of Massachusetts Amherst, Massachusetts USA
Araxy Terzian	Ithaca, New York USA
Yervant Terzian	Department of Astronomy Cornell University Ithaca, New York USA
Wes Traub	Harvard-Smithsonian Center for Astrophysics Cambridge, Massachusetts USA
Peter Usher	Department of Astronomy Pennsylvania State University University Park, Pennsylvania USA
Joe Veverka	Department of Astronomy Cornell University Ithaca, New York USA
Joy Veverka	Brooktondale, New York USA
Barbara Welther	Harvard-Smithsonian Center for Astrophysics Cambridge, Massachusetts USA
Fred Whipple	Harvard-Smithsonian Center for Astrophysics Cambridge, Massachusetts USA
Francis Wright	Harvard-Smithsonian Center for Astrophysics Cambridge, Massachusetts USA (*deceased*)

(*Photo by* Dennis di Cicco)

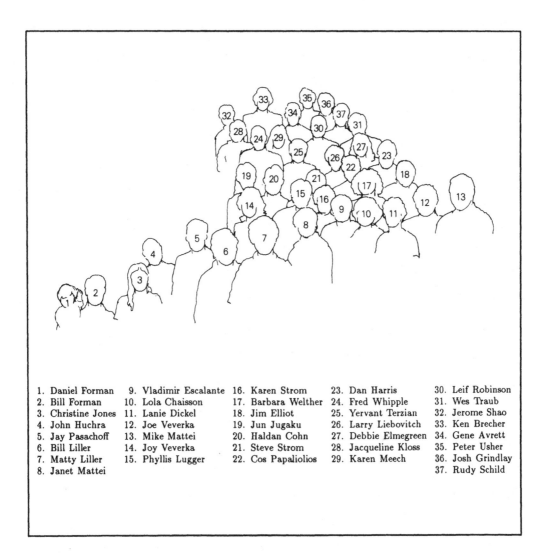

1. Daniel Forman	9. Vladimir Escalante	16. Karen Strom	23. Dan Harris	30. Leif Robinson
2. Bill Forman	10. Lola Chaisson	17. Barbara Welther	24. Fred Whipple	31. Wes Traub
3. Christine Jones	11. Lanie Dickel	18. Jim Elliot	25. Yervant Terzian	32. Jerome Shao
4. John Huchra	12. Joe Veverka	19. Jun Jugaku	26. Larry Liebovitch	33. Ken Brecher
5. Jay Pasachoff	13. Mike Mattei	20. Haldan Cohn	27. Debbie Elmegreen	34. Gene Avrett
6. Bill Liller	14. Joy Veverka	21. Steve Strom	28. Jacqueline Kloss	35. Peter Usher
7. Matty Liller	15. Phyllis Lugger	22. Cos Papaliolios	29. Karen Meech	36. Josh Grindlay
8. Janet Mattei				37. Rudy Schild

Joe Veverka (left) conversing with Yervant Terzian.

ASTEROIDS

J. Veverka
Center for Radiophysics and Space Research,
Cornell University, Ithaca, N.Y. 14853

INTRODUCTION

To some, asteroids are the "vermin of the skies";
to others, they are the Rosetta stone of the solar system.
The difference in perspective depends on whether one is
interested in taking long exposures of distant galaxies or
in understanding the steps by which the planets agglomerated
from smaller planetesimals some four and a half billion years
ago. Whatever one's perspective, it is undeniable that our
knowledge of these small bodies has increased significantly
during the past twenty years. This advance has resulted from
marked improvements in remote sensing techniques and from an
increased awareness that asteroids and their fragments, the
meteorites, preserve essential clues to some of the fundamental physical and chemical processes that were at work during

Figure 1. A preview of what asteroids look like?
The surface of the martian satellite Phobos imaged
by Viking Orbiter from a distance of 575 km. Hall,
the large crater at center right, is about 4 km
wide.

the initial phases of solar system history. While direct
exploration of the solar system has yet to take us to the
vicinity of an asteroid, we have had close looks at the
asteroids' first cousins, the small satellites of Mars and
of the outer planets. The perceived increased importance of
asteroids is reflected in NASA's planetary exploration
policy, which stipulates that all future missions to the
outer solar system, beginning with the Galileo mission to
Jupiter, shall investigate at least one asteroid en route.
Missions such as the planned Soviet-European venture VESTA,
which will combine in situ investigations with flybys of
several asteroids, are further proof of the increasing
interest in these small planets.

Twenty years ago, much of our information on asteroids was
restricted to knowledge of orbits and apparent magnitudes.
In several dozen cases, lightcurves were available from which
spin periods could be inferred. Not surprisingly, ideas
about asteroids in general and about their surfaces in
particular were naive. From this era come perceptions of
asteroids as irregularly shaped, wildly tumbling brickbats.

Fundamental gaps in our knowledge at that time included any
idea of what asteroid albedos are like, and therefore any
ability to infer true diameters. The situation is easy to
understand: from the apparent brightness of an asteroid we
cannot tell whether we are viewing a highly reflecting object
with a small cross section, rather than a low albedo object
with a large cross section. The difficulty arises because
asteroids are small, distant bodies which usually cannot be
resolved by earthbased telescopes unless special interferom-
etric techniques are employed. It is true that at the turn
of the century, Barnard (1902) made heroic efforts to measure
the apparent diameters of Ceres, Pallas, Juno, and Vesta,
using a filar micrometer, but his results were only approximately correct. Lacking firm values for asteroid diameters,
it was impossible to determine albedos; most investigators
resorted to simplifying assumptions such that most asteroids
had reflectances similar to that of the Moon.

Ideas concerning the nature of asteroid surfaces were equally
vague. For a small asteroid with a radius of 10 km, the
escape velocity is some 10 m/sec, and surface gravity only
10^{-3} g. Under such circumstances, what will be the effects
of prolonged meteoroid bombardment? Will the surface evolve
to look like that of the Moon, with readily recognizable
craters and a ubiquitous regolith of pulverized surface
material? Or, will the outcome be a regolith-free, pock-
marked surface of essentially bare rock?

ASTEROID DIAMETERS AND ALBEDOS
    In the late 1960's two independent methods were
developed to determine both the diameter and the albedo of an
asteroid. One of these, now referred to as the "radiometric

method," was pioneered by D. Allen, D. Matson, and D. Morrison, among others (e.g., Allen, 1970; Matson, 1971; Morrison, 1973). It resolves the ambiguity between cross section and albedo by making simultaneous observations of the asteroid's brightness at visual and infrared wavelengths. The brightness in visible light (after correction for distance factors such as the asteroid-Sun and the asteroid-Earth separations) depends on the projected cross section and on the albedo (A). At infrared wavelengths near 10-20 μm, one is detecting the asteroid's thermal radiation, which depends on the cross section and on (1 - A). Thus, a set of combined measurements yields both the albedo and the cross section, or diameter. The technique is readily applicable and has produced hundreds of determinations. Recently it has been augmented significantly by measurements made by IRAS, the Infrared Astronomy Satellite (Veeder, 1986).

The other method developed for determining albedos and diameters is based on polarization measurements. Following the pioneering work of Bernard Lyot in France during the 1920's (Lyot, 1929) it was well known that sunlight reflected from solar system surfaces is linearly polarized. By the 1960's it became evident that under certain circumstances the degree of linear polarization could be related to albedo: a dark surface producing relatively stronger polarization than a bright one (Dollfus, 1971; Veverka, 1971a). The technique has the advantage that polarization measurements also provide direct clues to the texture of the surface being observed.

My scientific relationship with William Liller developed around this question of using polarization measurements to determine the albedo and texture of asteroid surfaces. In the mid-1960's Liller had developed a polarimeter to measure polarization of interstellar dust, and was interested in making it available for other projects.

Figure 2 shows the polarization curve of asteroid 4 Vesta that I obtained during 1967-68, using the Liller polarimeter (Veverka, 1971b). It was evident from the results that Vesta has an intricately textured surface characteristic of a regolith (the negative branch at low phase angles is the evidence). From the slope of the positive branch (calibrated against laboratory data), one could estimate an albedo of 0.26, corresponding to a diameter of 515 km for Vesta. The values agree well with those obtained at about the same time by David Allen using the infrared technique outlined above (Allen, 1970), and have been substantiated by later results (Figure 3). In fact, during the past few years, it became possible to obtain resolved images of Vesta using speckle interferometry (Drummond et al., 1988).

We now know that asteroid albedos are far from all being alike or lunar-like. Most fall into two broad categories: very dark objects with albedos of 2-5%, and gray bodies with albedos between 10-25% (Morrison & Zellner, 1979). (For

Figure 2. Polarization curve of asteroid 4 Vesta obtained using the Liller polarimeter. The negative polarization values at phase angles below 20° suggest the presence of a regolith. The dashed curve is a smooth line drawn through the points, the solid lines represent a slope of 0.10 (%/deg) from which the albedo of the surface can be estimated (after Veverka, 1971b).

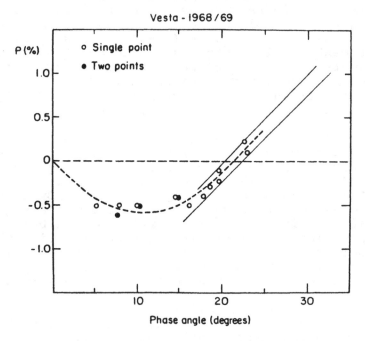

Figure 3. Early determinations of Vesta's diameter, compared with currently accepted values (TRIAD and Morrison & Zellner) The determination by Allen was made by the "radiometric" method; the "polarization curve" determination is based on the data in Figure 2.

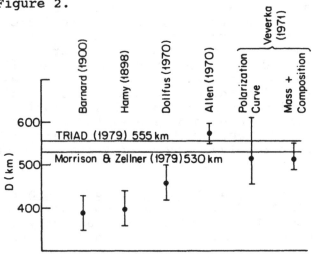

reference the lunar albedo is about 12%.) Asteroids as
reflective or more reflective than Vesta are rare. The
highest value currently known (0.48) is for asteroid 44 Nysa.

REGOLITHS ON ASTEROIDS
The early indications based on polarimetry that
asteroids have regoliths have since been substantiated by
other remote sensing techniques. Detailed thermal observations of asteroids as a function of solar phase angle imply
very low thermal inertias (rapid responses to changes in
insolation) consistent with the properties of porous regoliths (Morrison, 1976). Similar conclusions have been
reached on the basis of radar echoes from asteroids (Ostro et
al., 1985) and from investigation of photometric properties
of asteroid surfaces (Hapke, 1986). Asteroids typically show
a strong increase in brightness near zero phase, similar to
the behavior of our Moon when it reaches full phase. Such
strong "opposition surges" suggest a surface with the
intricate texture and high porosity of a regolith.

The original debate--can small, gravitationally weak bodies
have regoliths--was settled directly once and for all by
Mariner 9 when in 1971 it obtained the first detailed images
of Phobos and Deimos, the two small satellites of Mars
(Pollack et al., 1972; Veverka & Burns, 1980). Not only were
well-formed craters profusely evident, but photometric,
polarimetric, and thermal data all confirmed the presence of
regoliths. Perhaps the most dramatic proof came from eclipse
observations of Phobos made by the Mariner 9 Infrared Radi-
ometer, which measured surface temperatures as the satellite
emerged from the eclipse shadow of Mars (Gatley et al.,
1974). A remarkably rapid increase in temperature was
measured, proving that the surface was as good a thermal
insulator as the lunar regolith and certainly did not have
the thermal characteristics of solid rock!

When the Viking Orbiters followed up the study of Phobos and
Deimos in 1976 the existence of regoliths (including sizeable
blocks of ejecta) was confirmed directly by high resolution
images, which in some cases revealed detail as fine as 5-10
meters (Veverka and Burns, 1980). Can the Phobos/Deimos
results be extrapolated to asteroids? The available remote
sensing evidence (discussed above) suggests that the answer
is yes. But there are those who insist that the cases of
Phobos and Deimos are anomalous. They argue that since
Phobos and Deimos exist in the deep gravitational well of
Mars, most ejecta that leave the satellites cannot escape
Mars orbit and are therefore available to reimpact the
surfaces. How important a role this factor plays in determining the observed surface characteristics of the Martian
satellite remains uncertain.

It is noteworthy that increasingly sophisticated analyses of
asteroid thermal measurements continue to point to the

ubiquitous presence of low thermal inertia regoliths. Already, a decade ago, from a careful study of asteroid 433
Eros, Morrison (1976) was able to show that this particular
asteroid (an irregularly shaped body with dimensions 13 x 15
x 36 km) must be covered with a porous regolith.   More
recently, Lebofsky et al. (1985; 1986) have shown that
information on spin sense, pole orientation, as well as on
thermal regolith properties can be extracted from a careful
analysis of high quality thermal data (Figure 4).   Johnson et
al. (1983) have pioneered an analogous technique based on
polarization measurements in the thermal infrared. All these
studies indicate that at least the larger asteroids have
lunar-like regoliths, but there are suggestions from radar
results that this conclusion may not apply to some of the
smallest objects in the Earth-approaching population (Ostro
et al., 1983). Some of the tiny Apollos and Atens with
diameters of a few kilometers or less may not have well-
developed regoliths (Ostro, 1985).

Considerable progress has been made in understanding how even
small bodies can develop and retain regoliths (Housen et al.,
1979a, 1979b); Housen and Wilkening, 1980).   Fundamental to
this understanding is the laboratory demonstration by Stoffler et al. (1975) that ejecta velocities for impacts into
loose aggregate surfaces are significantly lower than for
corresponding impacts onto hard rock.   Thus, once a small
amount of ejecta builds up on the surface, a cushioning
effect is produced and there is a tendency for a larger
fraction of subsequent ejecta to be thrown out at less than
the escape velocity. Calculations by Kevin Housen predict
that even 10-20 km bodies could build up meters of regolith
(Figure 5, Veverka et al., 1986).

COMPOSITION OF ASTEROID SURFACES
         A close connection between meteorites and asteroids
has been suspected for almost two centuries. Even though the
quantitative details of where various meteorite samples have
come from and of the interrelationship of different meteorite
types with parent bodies in the asteroid belt are the
subjects of continued debate, most investigators agree that
most meteorites come from parent bodies in the asteroid belt.
Tremendous progress has been made during the past twenty
years in classifying asteroids into compositionally distinct
categories, and some progress has been made in connecting
these categories to available meteorite samples.

The first major advance in the classification of asteroids
occurred with the publication of a paper by Chapman, Morrison, & Zellner in 1975, in which they showed that on the
basis of color, albedo, and polarization properties, most
asteroids could be divided into two major broad categories:
the C and S objects (Chapman et al., 1975). Most investigators agree that the C-objects, which are characterized by
low albedos and relatively gray colors, are similar in

Figure 4. Two models fitted to observed infrared
fluxes for asteroid 10 Hygeia. A model which
assumes thermal characteristics similar to the
lunar regolith fits the measurements much better
than one in which the surface is assumed to behave
more like bare rock. (Modified after Lebofsky et
al., 1985).

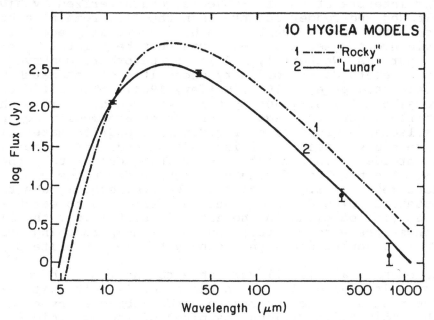

Figure 5. Regolith depth distributions on Phobos
(d = 21 km) and Deimos (d = 13 km) predicted by
Monte Carlo simulations of regolith production by
Kevin Housen. Qualitatively similar results are
predicted for small asteroids. (After Veverka et
al., 1986).

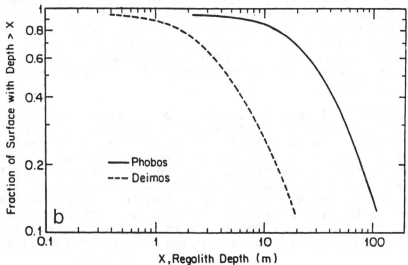

mineralogy to carbonaceous meteorites. An interesting subsequent development was the discovery by Lebofsky and his
associates that the spectra of some C-objects have an absorption feature near 3 microns attributable to water of
hydration (Lebofsky, 1978, 1980; Lebofsky et al., 1981). It
is known that certain types of carbonaceous meteorites (CI
and CM) contain 10-20% water associated with clay minerals.
The interpretation of S-objects, characterized by higher
albedos and redder spectra than the C's, remains a hotly
contested issue. On the one hand, several pieces of indirect
evidence suggest that S-objects must be related to ordinary
chondrites (Wetherill, 1985); on the other, observed spectra
of S-asteroids and ordinary chondrites do not match precisely
(Feierberg et al., 1982; Gaffey, 1984). The differences are
considered significant by some, while others argue that
poorly understood regolith processes are modifying spectral
reflectance properties of ordinary chondrite materials on S-
asteroid surfaces. The issue is a fundamental one: if S-
asteroids are not related to ordinary chondrites, then we
have the problem of finding a source for these most abundant meteorites. Worse still, we must explain why those S-
asteroids placed best to deliver fragments to earth--in
specific locations in the asteroid belt (e.g., the 3:1 resonance zone with Jupiter) and in earth-approaching orbits,
fail to contribute significantly to the meteorite population.

Chapman et al. (1975) were able to show that there is a
compositional gradation in the asteroid belt, with S-objects
dominating in the inner zone, and C-objects beyond about 2.7
AU. This "compositional mapping" of the asteroid belt, based
largely on albedo and color measurements, was gradually extended by other investigators, notably Gradie & Tedesco
(1982), and most recently by Tholen and his associates
(Tholen, 1984; Zellner et al., 1985). The Gradie & Tedesco
classification involves several additional asteroid categories, including the M and P asteroids. The M objects have
gradually become regarded as metallic--analogues of the iron/
nickel meteorites--and have been interpreted by some as
collisionally stripped cores of differentiated asteroids.
Metallic objects should have high radar cross section, and it
is significant that at least one metallic asteroid (1986 DA;
diameter = 2 km) has now been identified by both radar and
spectral reflectance techniques among the Earth-crossers
(Ostro, 1987).

The P-asteroids are believed to be carbonaceous and to be
related to the C's. The extensive "8-color survey" of Tholen
(1984) has been especially powerful in showing that there are
at least three types of carbonaceous asteroids (C, P, and D)
and that there is a gradual progression from C's to P's to
D's with increasing distance from the Sun (Figure 6; Hartmann, 1987; Cruikshank, 1986). The interrelationship is
unclear, in part because we do not have meteorite samples of
P or D composition. The D-asteroids, which predominate among
the Trojans, appear to have compositions dominated by complex

Figure 6. Distribution of major asteroid types with heliocentric distance. The S-asteroids dominate in the inner part of the asteroid belt. The C-asteroids peak near 3 AU, while their presumed relatives, the P's and D's, dominate farther out. After Cruikshank (1986).

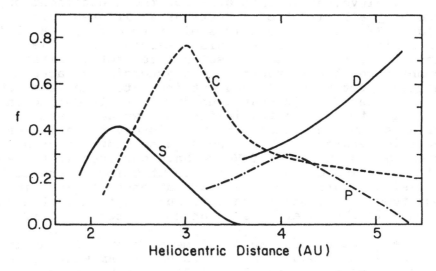

Figure 7. Comparison of the spectral reflectance curve of the Nuevo Laredo basaltic achondrite meteorite (solid curve) with that of asteroid 4 Vesta (points). From McCord et al., 1970). Subsequent data have confirmed the match. It appears that Vesta is covered with lava-like material that erupted from the asteroid's interior.

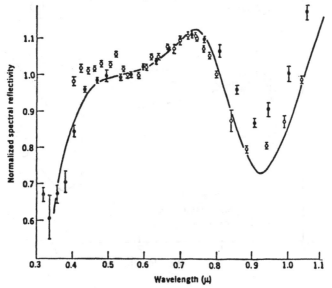

organic polymers and may be spectrally similar to comet
nuclei (Gradie & Veverka, 1980; Hartmann et al., 1982).

While most classification schemes utilize broad band color
data, considerable success has also been realized in interpreting measurements at higher spectral resolution. The
first achievement in this area was the identification by
McCord et al. (1970) of 4 Vesta as a differentiated object
(Figure 7) whose surface is similar to certain basaltic
achondrite meteorites (see also, Larson & Fink, 1975). This
finding was fundamental in several respects. First, it
demonstrated that meteorites and asteroid surfaces could be
matched spectrally. Second, it indicated that some objects
in the asteroid belt heated up sufficiently during their
evolution to melt on the inside and extrude lavas onto their
surfaces. Since Vesta's thermal history probably peaked some
4 billion years ago, the finding also showed that at least
some objects the size of Vesta have survived in the asteroid
belt for this length of time without being shattered by
collision. Currently, there are attempts to map mineral/
rock type variations on Vesta's surface from subtle spectral
changes in the asteroid's rotation lightcurve (Gaffey, 1983).

As already mentioned, spectral measurements in the infrared
have shown that some C-asteroids have 3-μm water of hydration
bands in their spectra (Figure 8). Further observations
indicate that some C-asteroids show the 3-μm absorption band,

Figure 8. Comparison of the spectrum of asteroid 1 Ceres
with spectra of three types of carbonaceous chondrites, two
of which contain abundant hydrated silicates. The measurements in the IR (crosses) indicate that water of hydration is
abundant in the surface of Ceres. (After Lebofsky, 1978).

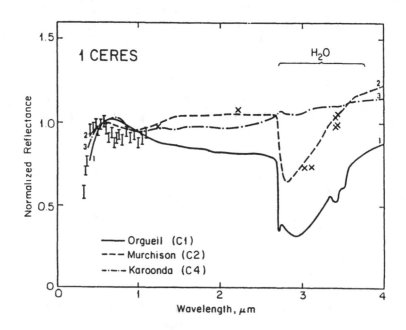

while others do not (Feierberg et al., 1985). The difference
could be related to the difference between CI/CM carbonaceous
meteorites which are water-rich, and CO and CV's which are
not. Recently, Cruikshank & Brown (1987) have announced the
detection of a 3.4-μm feature in the spectrum of at least one
dark carbonaceous asteroid, a finding which they interpret as
evidence of C-H bonds in an organic material.

A particularly intriguing result is the demonstration by
Cruikshank & Hartmann (1984), that some comparatively rare
asteroids with relatively high albedos and reddish spectra
can be understood as stony irons consisting of iron-nickel
metal and olivine. This conclusion, supported by the further
work of Bell et al. (1984), suggests that some asteroids are
the stripped-down cores of larger differentiated bodies (one
expects iron/olivine mixtures to occur at the core/mantle
boundary of a differentiated parent body). The concept of
collisionally stripped-down cores was developed by Clark
Chapman more than a decade ago (Chapman, 1976). In Chapman's
original view such stripped down cores were to be identified
with S-asteroids. While the debate of what S-asteroids
really are is unresolved, the existence of metallic asteroids
and of some with olivine-rich surfaces supports the concept
that significant removal of asteroid mantles by collisions
must have occurred in some instances.

THE STRUCTURE AND CONFIGURATION OF ASTEROIDS
While great strides have occurred in our understanding of the texture and composition of asteroid surfaces,
we still know very little about what asteroids look like,
except by inference. Until the first spacecraft exploration
of an asteroid is carried out, our best clues to what
asteroids are like come from the images of Phobos and Deimos
obtained by Mariner 9 and Viking (Figure 1, 9, 11). Phobos
and Deimos taught us a number of lessons which should guide
us in trying to understand asteroids. First: we must expect
to find unpredicted and unforeseen features and processes,
even on small bodies--grooves on Phobos (Thomas et al.,
1978); prominent downslope movement of regolith on Deimos
(Thomas & Veverka, 1980). Second: we must avoid thinking
that all small bodies are alike. Even though from afar
Phobos and Deimos are both dark, gray small bodies, in detail
they are quite different from one another. For some reason,
smaller Deimos seems to be more efficient at retaining fine-
grained ejecta (cf. Figure 11; Thomas & Veverka, 1980), even
though available calculations suggest the opposite (cf.
Figure 5; Veverka & Burns, 1980; Veverka et al., 1986). We
should expect much the same degree of diversity to obtain
among the first asteroids to be investigated by spacecraft.

Based on our experience with Phobos and Deimos, we should
expect asteroids to display heavily cratered surfaces with
well-developed regoliths made up of components ranging widely
in size from dust to large blocks of ejecta. Many smaller

<u>Figure 9</u>. The surface of Phobos is dominated by grooves, possibly formed by the severe impact which made Stickney, the largest crater on the satellite (Thomas <u>et al</u>., 1979). Thomas and Veverka (1979) have suggested that similar features will be found on some asteroids.

<u>Figure 10</u>. Schematic diagram showing the results of four categories of target damage at progressively higher energies. After Thomas and Veverka (1979).

IMPACT   ENERGY

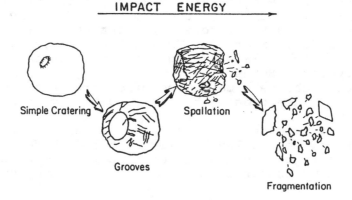

Simple Cratering   Spallation

Grooves

Fragmentation

asteroids should have irregular shapes due to long histories
of impacts; some should show evidence of nearly catastrophic
collisions. A major discovery by Viking was the existence of
a system of long, linear, negative-relief features associated
with Stickney, the largest crater on Phobos. Thomas,
Veverka, Bloom, & Duxbury showed that these grooves were
generated at the time of the impact that produced Stickney
(Thomas et al., 1978). Experiments on laboratory-sized
targets by Fujiwara and his associates in Japan support the
hypothesis that Stickney and the grooves can both result from
a single nearly-catastrophic impact long ago (e.g., Fujiwara,
1986). Thomas & Veverka (1979) have proposed that groove-
like features should occur on some asteroids: those which in
their current configuration experienced an impact close to,
but not exceeding about 10% of the energy necessary for
disruption. Laboratory simulations by Gault & Wedekind
(1969), as well as by Fujiwara (1986) and others, show that
more severe impacts lead to catastrophic disruption of the
original body, usually by the spallation of the mantle layers
and the preservation of an inner core (Figure 10).

There have been suggestions that as a result of continued
collisions and impacts some asteroids today are no more than
"rubble piles," gravitationally-bound aggregates of collisionally fractured debris (e.g., Farinella et al., 1981,
1982). We do not know whether any such hypothetical rubble
piles really exist in the solar system. One characteristic

Figure 11. A high-sun (low phase angle) view of
Deimos. Patches of brighter material are conspicuous, as is evidence of apparently loose surface material filling depressions. See Thomas and
Veverka (1980), for a detailed discussion of evidence of downslope movement of regolith on Deimos.

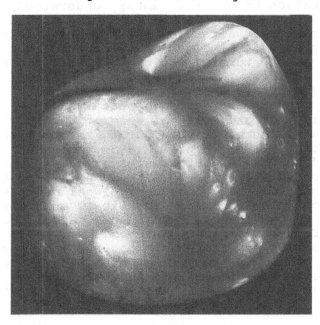

would be an unusually low mean density, due to the presence of considerable void space (the alternative explanation of low densities, the presence of ice, can be excluded in much of the asteroid belt on thermal stability grounds).

It is worth stressing that only in the case of the three asteroids (Ceres, Pallas, and Vesta) has it proved possible to determine masses by traditional perturbation methods (Schubart & Matson, 1979). For most, masses remain unknown and can only be determined from spacecraft encounters. Mass determinations that lead to precise measurements of average densities will provide important constraints on models of asteroid interiors. Consequently, mass determinations are prime objectives of all planned spacecraft investigations of asteroids.

Today we know nothing directly about the internal structure of any asteroid. We can draw inferences from meteorites, and speculate from knowledge of surface composition, and of course, we can model. At present, such modelling is ill-constrained, since we lack detailed knowledge of many fundamental parameters. For example, if we wish to follow the collisional evolution of a particular body, we need to know the nature and time evolution of the cratering population. At present, we barely know these data in the vicinity of the Earth-Moon system, let alone elsewhere. Our modelling is especially hampered because we cannot draw on any useful experience to tell us how hundred-kilometer objects respond to catastrophic hypervelocity impacts. Our theories are not sophisticated enough to provide definitive answers; our best laboratory simulations fail in scale by many orders of magnitude. We should not be too dogmatic about what asteroids are like; we need to explore and learn.

One indication of our rudimentary understanding of asteroids is the persistent popularity of the suggestion that binary, or even multiple, asteroids exist (see Van Flandern et al., 1979, for a review). The idea is not a new one, having been proposed at the turn of the century to explain the light curve of 433 Eros (André, 1901). The possibility is difficult to disprove absolutely a priori and in the hand of dexterous devotees provides a ready (and usually untestable) explanation for all sorts of enigmatic observations. Dozens of papers have been written during the past ten years (e.g., Sichao et al., 1981; Binzel, 1985), and even more talks presented on the subject, but never has it been proven that any asteroid is a binary. Gehrels et al. (1987), in a recent review, conclude both on observational grounds and on revised theoretical considerations that it is unlikely that any asteroid has a sizeable "permanent" satellite. Unfortunately, the fantasy of binary asteroids will persist until it can be put to rest by irrefutable observational evidence. In cases of particular individual asteroids, future missions should serve the purpose (for example, the alleged binary 46 Hestia may be explored by the Soviet/French mission VESTA);

for the asteroid population at large, direct imaging from Space Telescope should help.

COMETS AND ASTEROIDS

A question of perennial interest is how does one tell a dead comet from an asteroid? It is clear that short-period comets should gradually lose their volatiles, but the end state, if indeed there is a single end state, remains unknown. It is possible that one is left with a sizeable body whose surface has the characteristics recently documented for comet nuclei--that is, dark and spectrally reddish, similar to D-asteroids. Another possibility is that comets disintegrate completely into a meteor stream as they evolve. That meteor streams are associated with comets has been established on the basis of orbital evidence for more than a century. This connection has been confirmed vividly by the recent IRAS discovery of dust bands associated with individual period comets (e.g., Tempel 2) (Sykes et al., 1986).

The realization that some comets may "turn off" has traditionally led to suggestions that one could tell "dead" or "inactive" comets on the basis of their orbits. Short period comets and main belt asteroids differ on average in their orbital properties, more elongated orbits with aphelia near Jupiter being characteristic of comets. Some asteroid-like objects in similar orbits, especially those with dark, spectrally reddish surfaces, could be dead comets.

The search for dead comets brings me back to William Liller, who has maintained a lifelong interest in comets and asteroids. When in 1968, the unusual Apollo asteroid 1566 Icarus made a close passage to Earth, Liller and I carried out a series of observations, partly in the hope of shedding some light on whether Icarus could be a dead comet nucleus (Veverka & Liller, 1969). The remote sensing data obtained for Icarus showed rather clearly that in spite of its very elongated and inclined orbit, Icarus is a rocky object, probably related in composition to ordinary chondrite meteorites, and not a dead comet nucleus. While the UBV colors measured for Icarus do not correspond precisely to those of S-asteroids, available spectral data (McFadden et al., 1984) demonstrate a close similarity between Icarus and two other Apollo asteroids--1685 Toro and 1862 Apollo. McFadden et al. (1985) show that 1862 Apollo is spectrally similar to some LL4 ordinary chondrites. In passing, it should be noted that no evidence of cometary activity associated with Icarus has ever been reported.

Currently, there are two popular "dead comet" candidates among the asteroids. One is 2201 Oljato, a bright (albedo estimated at 40-50%), 1.5 km diameter object in an unusual orbit which takes it from 0.63 AU at perihelion to 3.72 AU at aphelion. Russell et al. (1984) reported the detection by

Venus Orbiter instruments of a disturbance in the interplanetary magnetic field which they related to material that
could have been ejected from Oljato. McFadden et al. (1984)
observed a unique reflectance curve--unlike that of any other
asteroid--which includes emission in the violet region of the
spectrum. The emissions do not correspond to well-known
cometary features such as CN or C_2, and remain unidentified.

The second candidate is asteroid 1983TB, also known as 3200
Phaethon (Davies, 1985). Discovered by IRAS and shown to
have an orbit associated with the Geminid meteors, the
asteroid has been considered a prime dead-comet candidate by
some on the basis of its connection with the meteor stream.
Phaethon is now known to be an irregularly shaped body with
an average diameter of 3-5 km, and an albedo of about 10%.
The albedo and color of Phaethon are S-like, and certainly
not D-like (Davies, 1985). A high-resolution spectral study
between 3500 and 6500 Å of Phaethon has been carried out by
Cochran & Barker (1984), who found no evidence of cometary
emissions or activity. Their conclusions are worth quoting:

> "... we can find no direct evidence that
> 1983TB is a 'dead' comet. However, had
> we found some cometary emissions, it
> would not have showed that a comet had
> evolved into an asteroid but that a comet
> had been misclassified as an asteroid."

As the quote illustrates, one problem with trying to identify
"dead comets" is that one may be dealing with an ill-posed
question which is not amenable to answer by current remote
sensing techniques.

THE FUTURE
The first direct observation of an asteroid by
spacecraft is still 5-10 years away. The Galileo mission to
Jupiter, had it been launched as scheduled in May 1986, would
have flown by the large S-type asteroid Amphitrite in
December 1986. The current schedule calls for a launch in
late 1989, followed by two asteroid flybys, the first of
which is likely to be that of the 16 km S-type asteroid 951
Gaspra in October 1991. There are plans for a joint Soviet-
French-ESA mission called VESTA, which might include flybys
and penetrator studies of several asteroids, including the P-
asteroid 46 Hestia. There are less definite plans for NASA
missions to the outer solar system (CRAF and Cassini) to
include asteroid flybys. As a result of these efforts, some
of which are still 5-10 years off, we will finally learn what
asteroids look like. By and large, we expect to find that
extrapolations made from our knowledge of small satellites
are valid. Eventually, when we get to study very small
bodies, less than a kilometer across, we should not be
surprised to find that such tiny objects might look more like
large rock fragments than minuscule planetary bodies.

Already, there are suggestions in radar investigations that some small Earth-approaching asteroids may have more rugged (less regolith-covered) surfaces than is the case for larger main belt objects. And perhaps one day, if we explore enough, we will even find an asteroid that everyone will agree must be a "dead comet."

***

Question (J. Elliot): How strong is the evidence for binary asteroids from lightcurve data?

Answer: The suggestion that some asteroid lightcurves can be interpreted in terms of a binary model is not new and goes back to Andre's unsuccessful attempt at the beginning of this century to explain the then seemingly unusual lightcurve of 433 Eros. The basic problem is one of uniqueness. For some asteroid lightcurves it is not difficult to construct a binary model which fits some of the observations. My problem with such exercises is that they are highly non-unique, a fact which is well known to the modellers. Also well known to the modellers is the fact that one can always obtain an equally good, or even better fit, by modelling the asteroid as an irregularly shaped object with (in some cases) albedo markings. Personally, I have little patience with claims that asteroids such as 46 Hestia must be binaries because of their lightcurves. What I want to know is why the more prosaic explanation in terms of irregular shape and albedo markings cannot do the job. I have yet to hear a reasonable answer. Thus, I come down on the side of those who would say that a lot of this binary modelling is an exercise in pedantry rather than science.

Question (W. Liller): What characteristics should one look for when trying to identify a "dead comet"?

Answer: We really don't know. It is true that our knowledge of comet nuclei has been improving steadily, but we are far from being able to predict what a comet that has lost its volatiles will look like. Currently, the best guess is that we should be looking for a very dark body, smaller than 10 km in radius, with a reddish D-type spectrum. But it is very possible that there are objects in the asteroid belt that match this description, but which have never been "comets." Another clue is the object's orbit. We would expect dead comets to be in orbits that are similar to, or at least derivable from, those of short period comets. Thus, small, dark-reddish asteroids in elongated orbits with aphelia near Jupiter are good "dead comet" candidates.

Question (J. Grindlay): Could you or Cos Papaliolios comment
on the current possibilities and status of speckle observations of asteroids? In particular, have there been speckle
measurements of 4 Vesta, and if so, how do the speckle data
compare with inferences from analyses of lightcurves?

Answer (C. Papaliolios): Speckle imaging can show brightness
variations of asteroids at about 0.1 arcsec resolution.
There are some preliminary results on Vesta shown in the June
1987 issue of Sky and Telescope obtained by the Arizona group
of Drummond, Eckart, and Hege.

Answer (J. Veverka): The speckle data on Vesta, including a
series of reconstructed images, are being published in Icarus
(1988, 73, 1-14). Drummond et al. interpret their data to
mean that Vesta has an irregular shape (triaxial diameters of
584 ± 16, 531 ± 11, and 467 ± 12 km) and a spotted surface.
The mean diameter of 525 km agrees well with the results in
Figure 3. Currently there remains a discrepancy between the
inference that Vesta is both irregularly shaped and spotted,
and some analyses of the lightcurve in terms of a spherical
asteroid with a spotted surface. Connected with this is a
debate over Vesta's true spin period. The speckle data suggest strongly that Vesta spins once every 5h 20.5m, whereas
some interpretations of the lightcurve yield a period twice
as long. Speckle data by the Arizona group for other asteroids (including 433 Eros, 511 Davida, and 532 Herculina) can
be found in preceding issues of Icarus.

REFERENCES

Allen, D. A. (1970). The infrared diameter of Vesta. Nature
        227, 158-159.
André, C. (1901). Sur le système formé par la planète double
        (433) Eros. Astron. Nach. 155, 27-30.
Barnard, E. E. (1900). The diameter of Vesta. Monthly Not.
        Roy. Astron. Soc. 61, 68.
Barnard, E. E. (1902). On the dimensions of the planets and
        satellites. Astron. Nach. 157, 260-261.
Bell, J. F., Gaffey, M. J., & Hawke, B. R. (1984). Spectro-
        scopic identification of probable pallasite parent
        bodies. Meteoritics 19, 187-188.
Binzel, R. P. (1985). Is 1220 Crocus a precessing, binary
        asteroid? Icarus 63, 99-108.
Chapman, C. R. (1976). Asteroids as meteorite parent bodies:
        The astronomical perspective. Geochim. Cosmochim.
        Acta 40, 701-719.
Chapman, C. R., Morrison, D., & Zellner, B. (1975). Surface
        properties of asteroids: A synthesis of polarimet-
        ry, radiometry, and spectrophotometry. Icarus 25,
        104-130.
Cochran, A. L., & Barker, E. S. (1984). Minor planet 1983TB:
        A dead comet? Icarus 59, 296-300.
Cruikshank, D. P. (1986). Dark material in the solar system.
        COSPAR Invited Paper.
Cruikshank, D. P., & Brown, R. H. (1987). Organic matter on
        asteroid 130 Elektra. Science 238, 183-184.
Cruikshank, D. P., & Hartmann, W. K. (1984). The meteorite-
        asteroid connection: Two olivine-rich asteroids.
        Science 223, 281-282.
Davies, J. K. (1985). Is 3200 Phaethon a dead comet? Sky
        and Telescope 70, 317-318.
Dollfus, A. (1970). Diamètres des planètes et satellites.
        In Surfaces and Interiors of Planets and Satellites
        (A. Dollfus, ed.), Academic Press, N.Y.
Dollfus, A. (1971). Physical studies of asteroids by
        polarization of their light. In Physical Studies
        of Minor Planets (T. Gehrels, ed.), NASA SP-267,
        Washington, D.C.
Drummond, J., Eckart, A., & Hege, E. K. (1988). Speckle
        interferometry of asteroids. IV. Reconstructed
        images of Vesta. Icarus 73, 1-14.
Farinella, P., Paolicchi, P., & Zappalà, V. (1982). The
        asteroids as outcomes of catastrophic collisions.
        Icarus 52, 409-433.
Farinella, P., Paolicchi, P., Tedesco, E. F., & Zappalà, V.
        (1981). Triaxial equilibrium ellipsoids among the
        asteroids? Icarus 46, 114-123.
Feierberg, M. A., Larson, H. P., & Chapman, C. R. (1982).
        Spectroscopic evidence for undifferentiated S-type
        asteroids. Astrophys. J. 257, 361-372.
Feierberg, M. A., Lebofsky, L. A., & Tholen, D. J. (1985).
        The nature of C-class asteroids from 3 μm spectro-
        photometry. Icarus 63, 183-191.

Fujiwara, A. (1986). Results obtained by laboratory simula-
        tions of catastrophic impact. Mem. Soc. Astron.
        Italiano 57, 47-64.
Gaffey, M. J. (1983). The asteroid 4 Vesta: Rotational
        spectral variations, surface material heterogene-
        ity, and implications for the origin of basaltic
        achondrites. Lunar Planet. Sci. Conf. (Abstract)
        XIV, 231-232.
Gaffey, M. J. (1984). Rotational spectral variations of
        asteroid (8) Flora. Icarus 60, 83-114.
Gatley, I., Kieffer, H., Miner, E., & Neugebauer, G. (1974).
        Infrared observations of Phobos from Mariner 9.
        Astrophys. J. 190, 497-503.
Gault, D. E., & Wedekind, J. (1969). The destruction of
        tektites by micrometeoroid impact. J. Geophys.
        Res. 74, 6780-6794.
Gehrels, T., Drummond, J. D., & Levenson, N. A. (1987). The
        absence of satellites of asteroids. Icarus 70,
        257-263.
Gradie, J., & Tedesco, E. (1982). Compositional structure of
        the asteroid belt. Science 216, 1405-1407.
Gradie, J., & Veverka, J. (1980). The composition of Trojan
        asteroids. Nature 283, 840-842.
Hamy, M. (1898). Bull. Astron. 16, 256.
Hapke, B. W. (1986). Bidirectional reflectance spectroscopy.
        4. The extinction coefficient and the opposition
        effect. Icarus 67, 264-280.
Hartmann, W. K. (1987). A satellite-asteroid mystery and a
        possible early flux of scattered C-Class asteroids.
        Icarus 71, 57-68.
Hartmann, W. K., Cruikshank, D. P., & Degewij, J. (1982).
        Remote comets and related bodies: VJHK colorimetry.
        Icarus 52, 377-408.
Housen, K. R., & Wilkening, L. L. (1982). Regoliths on small
        bodies in the solar system. Ann. Rev. Earth
        Planet. Sci. 10, 355-376.
Housen, K. R., Wilkening, L. L., Chapman, C. R., & Greenberg,
        R. (1979a). Asteroidal regoliths. Icarus 39,
        317-351.
Housen, K. R., Wilkening, L. L., Chapman, C. R., & Greenberg,
        R. J. (1979b). Regolith development and evolution
        on asteroids and the moon. In Asteroids (T.
        Gehrels, ed.), U. of Arizona Press, Tucson.
Johnson, P. E., Kemp, J. C., Lebofsky, M. J., & Rieke, G. H.
        (1983). 10 μm polarimetry of Ceres. Icarus 56,
        381-392.
Larson, H. P., & Fink, U. (1975). Infrared spectral observa-
        tions of asteroid 4 Vesta. Icarus 26, 420-427.
Lebofsky, L. A. (1978). Asteroid 1 Ceres: Evidence for water
        of hydration. M.N.R.A.S. 182, 17-21.
Lebofsky, L. A. (1980). Infrared reflectance spectra of
        asteroids. A search for water of hydration.
        Astron. J. 85, 573-585.

Lebofsky, L. A., Feierberg, M. A., Tokunaga, A. T., Larson,
    H. P., & Johnson, J. K. (1981). The 1.4-4.2 μm
    spectrum of asteroid 1 Ceres. Icarus 48, 433-459.
Lebofsky, L. A., Sykes, M. V., Nolt, I. G., Radostitz, J. V.,
    Veeder, G. J.,Matson, D. L., Ade, P.A.R., Griffin,
    M. J., Gear, W. K., & Robson, E. I. (1985).
    Submillimeter observations of the asteroid 10
    Hygiea. Icarus 63, 192-200.
Lebofsky, L. A., Sykes, M. V., Tedesco, E. F., Veeder, G. J.,
    Matson, D. L., Brown, R. H., Gradie, J. C., Feier-
    berg, M. A., & Rudy, R. J. (1986). A refined
    "standard thermal model for asteroids based on
    observations of 1 Ceres and 2 Pallas. Icarus 68,
    239-251.
Lyot, B. (1929). Recherches sur la polarisation de la
    lumière des planètes et de quelques substances
    terrestres. Doctoral Thesis, University of Paris.
    NASA Tech. Translation F-187, 1964.
Matson, D. L. (1971). Infrared observations of asteroids.
    In Physical Studies of Minor Planets (T. Gehrels,
    ed.), NASA SP-267, Washington, D. C.
McCord, T. B., Adams, J. B., & Johnson, T. V. (1970).
    Asteroid Vesta: Spectral reflectivity and composi-
    tional implications. Science 168, 1445-1447.
McFadden, L. A., Gaffey, M. J., & McCord, T. B. (1984).
    Mineralogical-petrological characterization of
    near-earth asteroids. Icarus 59, 25-40.
McFadden, L. A., Gaffey, M. J., & McCord, T. B. (1985).
    Near-earth asteroids: Possible sources from
    reflectance spectroscopy. Science 229, 160-163.
Morrison, D. (1973). Determination of radii of satellites
    and asteroids from radiometry and photometry.
    Icarus 19, 1-14.
Morrison, D. (1976). The diameter and thermal inertia of 433
    Eros. Icarus 28, 125-132.
Morrison, D., and Zellner, B. (1979). Polarimetry and radi-
    ometry of the asteroids. In Asteroids (T. Gehrels,
    ed.), U. of Arizona Press, Tucson.
Ostro, S. J. (1985). Radar properties of near-earth aster-
    oids. Bull. Amer. Astron. Soc. 17, 729-730.
Ostro, S. J. (1987). Personal communication.
Ostro, S. J., Campbell, D. B., & Shapiro, I. I. (1983).
    Radar observations of asteroid 1685 Toro. Astron.
    J. 88, 565-576.
Ostro, S. J., Campbell, D. B., & Shapiro, I. I. (1985).
    Mainbelt asteroids: Dual-polarization radar
    observations. Science 229, 442-446.
Pollack, J. B., Veverka, J., Noland, M., Sagan, C., Hartmann,
    W. K., Duxbury, T. C., Born, G. H., Milton, D. J.,
    & Smith, B. A. (1972). Mariner 9 television
    observations of Phobos and Deimos. J. Geophys.
    Res. 78, 4313-4326.
Russell, C. T., Aroian, R., Arghavani, M., & Nock, K. (1984).
    Interplanetary magnetic field enhancements and
    their association with asteroid 2201 Oljato.
    Science 226, 43-45.

Schubart, J., & Matson, D. (1979). Masses and densities of
          asteroids. In Asteroids (T. Gehrels, ed.), U. of
          Arizona Press, Tucson.
Sichao, W., Yuezhen, W., Mengxian, B., Liwu, D., & Sufang, W.
          (1981). A possible satellite of 9 Metis. Icarus
          46, 285-287.
Stoffler, D., Gault, D. E., Wedekind, J., & Polkowski, G.
          (1975). Experimental hypervelocity impact into
          quartz sand: Distribution and shock metamorphism of
          ejecta. J. Geophys. Res. 80, 4062-4077.
Sykes, M. V., Lebofsky, L. A., Hunten, D. M., & Low, F.
          (1986). The discovery of dust trails in the orbits
          of periodic comets. Science 232, 1115-1117.
Tholen, D. (1984). Asteroid taxonomy from cluster analysis
          of photometry. Ph.D. Dissertation, University of
          Arizona, Tucson.
Thomas, P., & Veverka, J. (1979). Grooves on asteroids: A
          prediction. Icarus 40, 395-406.
Thomas, P., & Veverka, J. (1980). Downslope movement of
          material on Deimos. Icarus 42, 234-250.
Thomas, P., Veverka, J., & Duxbury, T. C. (1978). Origin of
          grooves on Phobos. Nature 273, 202-204.
Thomas, P., Veverka, J., Bloom, A., & Duxbury, T. (1979).
          Grooves on Phobos: Their distribution, morphology,
          and possible origin. J. Geophys. Res. 84, 8457-
          8477.
TRIAD (1979). The Tucson Revised Index of Asteroid Data. In
          Asteroids (T. Gehrels, ed.), U. of Arizona Press,
          Tucson.
Van Flandern, T. C., Tedesco, E. F., & Binzel, R. P. (1979).
          Satellites of asteroids. In Asteroids (T. Gehrels,
          ed.), U. of Arizona Press, Tucson.
Veeder, G. J. (1986). Characteristics of the asteroid data.
          In IRAS Asteroid and Comet Survey (D. Matson, ed.),
          pp. 2-1 to 2-53. JPL Document No. 3698.
Veverka, J. (1971a). Asteroid polarimetry: A progress
          report. In Physical Studies of Minor Planets (T.
          Gehrels, ed.), NASA SP-267, Washington, D.C.
Veverka, J. (1971b). The polarization curve and absolute
          diameter of Vesta. Icarus 15, 11-17.
Veverka, J., & Burns, J. (1980). The moons of Mars. Ann.
          Rev. Earth Planet. Sci. 8, 527-558.
Veverka, J., & Liller, W. (1969). Observations of Icarus:
          1968. Icarus 10, 441-444.
Veverka, J., Thomas, P., Johnson, T. V., Matson, D., &
          Housen, K. (1986). The physical characteristics of
          satellite surfaces. In Satellites (J. Burns and M.
          Matthews, eds.), U. of Arizona Press, Tucson.
Wetherill, G. W. (1985). Asteroidal sources of ordinary
          chondrites. Meteoritics 20, 1-22.
Zellner, B., Tholen, D. J., & Tedesco, E. F. (1985). The
          eight-color asteroid survey. Icarus 61, 355-416.

(*Photo by* Marjorie Nichols)

Jim Elliot (centre) conversing with Larry Liebovitch, Jay Pasachoff, and Christine Jones.

SIXTEEN YEARS OF STELLAR OCCULTATIONS

Jim Elliot

Department of Earth, Atmospheric, and Planetary Sciences
Department of Physics,
Massachusetts Institute of Technology,
Cambridge, Massachusetts 02139, USA

Abstract. Stellar occultations by solar system bodies have emerged from an infrequent "astronomical happening" in the 1950's and 60's, to a routine tool used by a growing number of planetary astronomers in the 1970's and 80's. The increasing popularity of the technique is due to (i) the 10,000-fold improvement in spatial resolution that it affords over ordinary ground-based methods for probing the outer solar system and (ii) several technical advancements that have allowed occultations to be observed more frequently and with continually improving signal-to-noise ratios. In addition to large telescopes at fixed locations, the observing platforms have included portable telescopes on the ground, the Voyager spacecraft, and NASA's Kuiper Airborne Observatory. Major results from the occultation technique include: the discovery of thermal waves in the Martian atmosphere; the discovery and comprehensive study of the Uranian ring system; the accurate measurement of the sizes and shapes of several asteroids; the discovery of waves and other features in the Saturnian rings; the probing of the upper atmospheres of the gas giant planets; the discovery of the Neptune "arcs"; and most recently, the detection of an atmosphere surrounding Pluto. In the future, further development of the occultation technique can be expected with observations from the Hubble Space Telescope and the use of CCD's for occultation predictions and observations.

1 OCCULTATIONS FOR FUN AND PROFIT

Why observe stellar occultations? For me, the main reason for observing them is that the best occultations yield *10,000 times better spatial resolution* for probing bodies in the outer solar system than other Earth-based techniques. Even with speckle interferometry, a telescope *several kilometers* in diameter would be required to achieve the equivalent spatial resolution at a wavelength of 2 μm! The reason for the high spatial resolution is that we are not imaging distant bodies, with the attendant problems with atmospheric "seeing." Rather, we are probing them with starlight that is transmitted, refracted, and diffracted by the distant bodies before it has been disturbed by the Earth's atmosphere. Hence, the spatial resolution of our occultation observations is limited by some combination of (i) Fresnel diffraction, (ii) the angular diameter of the occulted star, and (iii) the signal-to-noise ratio of the data. Seeing and Fraunhofer diffraction effects by the telescope used to collect the starlight do not affect the spatial resolution for probing the occulting body.

Observation of a stellar occultation simply amounts to recording the starlight as a function of time, while the relative motion of the Earth and the occulting body produces a "line scan" of stellar intensity as a function of position on the occulting body. If occultation observations can be carried out simultaneously from more than one telescope, we then have multiple line scans across the occulting body, which give us more complete information.

Depending on the physical characteristics of the occulting body, the line scan can be a trace of optical depth through ring material, a probe of the differential refraction produced by an atmosphere, or just an "on-off" event that gives us the length of a chord across an asteroid or small satellite. Relating the intensity trace to an accurate time base is critical, since the time tag on each stellar intensity point is directly related (through an ephemeris) to a position on the occulting body through which the starlight passed. A complete discussion of observing and analysis methods is far beyond the scope of this review, and the reader is referred to Elliot (1979), Elliot & Nicholson (1984), and the references therein for more information.

The occultation technique has proven its value through the discoveries of the Uranian rings, the Neptune ring-arcs, and thermal waves in the Martian upper atmosphere. From a single observatory we measure a line scan of the transmitted starlight across a single chord through a body; repeated and/or multiple observations of occultations allow us to build a more complete picture, and, in addition, to study changes over time of what we measure. As an example, we have constructed a highly sophisticated model for the orbits of the Uranian rings by combining the occultation observations obtained over the last decade.

Since the high-speed photometric techniques used to observe stellar occultations have been available since the 1950's, and in view of the high spatial resolution obtainable with occultations, you may be surprised at what is shown in Table I: only six stellar occultations were observed during 25 years! Following the time interval shown in Table I, at least one stellar occultation has been observed each year, with the yearly average being about 2-3.

What caused the abrupt increase in the popularity of observing occultations? The answer to this question lies in effects of (i) several technical developments (more occultation predictions, better detectors, and mobile observatories), (ii) an increased appreciation of what can be learned from occultation data (especially in the study of planetary rings), and (iii) the parallel development of planetary ring theory, which greatly enhanced the significance of what we learned from the observations.

The story of all developments in the field of occultations would require a week-long conference and would fill a large volume—perhaps several volumes. Hence this review will be limited to the developments in the field with which I have been most closely associated. It emphasizes the work that Bill Liller was involved in, since this is, after all, a volume that celebrates not only Bill's sixtieth birthday, but his research achievements in fields ranging from "asteroids to quasars" and his influence on the many people that he encountered along the way.

TABLE I

STELLAR OCCULTATIONS OBSERVED* THROUGH MARCH 1977

Planet	Star	Date	Years After Prior Event
Jupiter	σ Arietis	1952 November 20	...
Venus	Regulus	1959 July 7	7
Neptune	BD–17°4399	1968 April 7	9
Jupiter	β Scorpii	1971 May 13	3
Mars	ε Geminorum	1976 April 8	5
Uranus	SAO 158687	1977 March 10	1

*photoelectric observations; for references, see Elliot (1979). To the best of my knowledge these are also the only stellar occultations by major planets predicted by Gordon Taylor for this time interval.

2   BEGINNING WITH β SCORPII ...

My introduction to stellar occultations occurred in early 1971, when Joe Veverka, of Cornell University, asked if I would join his team for observing the β Scorpii occultation by Jupiter, which would occur in May of that year. Since our observations would be made from Boyden Observatory in South Africa, we would have to transport our equipment to the site as extra luggage. Joe had already recruited Bill Liller to build a portable, three-channel photometer and the pulse counters; my job would be to build the data recording system that would put the photon pulses onto audio tape, synchronously with radio time signals. The fourth member of our team was Larry Wasserman, then a graduate student at Cornell. This expedition appealed to me as an exciting opportunity, although it would be a diversion from completing my degree. Little did I suspect that the adventure would continue for years!

The goal of our observations was to determine the scale height of Jupiter's atmosphere by measuring the rate at which the star would dim through the process of *differential refraction*. This is illustrated in Figure 1, where we see light from a distant star incident as parallel rays on the planetary atmosphere from the left. Since the density gradient in the atmosphere is exponential, each successive ray is refracted by the atmosphere more than the previous one, as shown in the figure. Even though an insignificant amount of light is lost to extinction, to a distant observer, the star appears to dim through differential refraction,. The resulting light curve is shown at the right of Figure 1, where the rate of dimming is

Figure 1 Stellar occultation geometry and optics. In this ray diagram, the light from the occulted star is incident from the left and is refracted by the density gradient in the planetary atmosphere. Since the density gradient becomes progressively greater, deeper into the atmosphere, the rays are spread by an ever increasing amount by the process of differential refraction. As the Earth travels through the pattern of refracted starlight, the stellar intensity decreases as shown by the graph at the right-hand side of the figure. The equation describing the starlight intensity as a function of time for the isothermal atmosphere illustrated here is given by Baum & Code (1953). (after Elliot 1979).

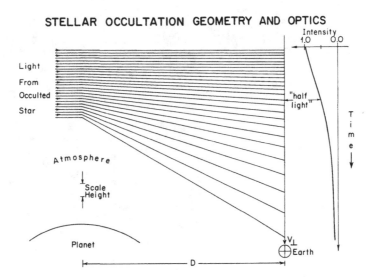

STELLAR OCCULTATION GEOMETRY AND OPTICS

determined by the scale height, H, of Jupiter's atmosphere. This is related to the temperature, T, mean molecular weight, μ, gas constant, R, and local gravity, g, through the equation, $H = RT/\mu g$. Since the local gravity is usually well known, with prior knowledge of either the temperature or the mean molecular weight, we can determine the other from our measured scale height.

Figure 1 shows the light curve that would be observed for an isothermal atmosphere, but during an occultation of BD–17°4399 by Neptune three years earlier, Ken Freeman and his colleagues at Mount Stromlo had observed not a smooth light curve, but one that exhibited bright flashes (or "spikes"; Freeman & Lyngå 1970). Joe wanted to see if the spikes would be produced by Jupiter's atmosphere, so he wanted our data recorded at 10 ms time resolution. For redundancy, he wanted the photometer to have three channels, and later we learned that by observing the spikes at separated wavelengths we could infer the helium abundance in Jupiter's atmosphere through the differences in refractivity between hydrogen and helium (Brinkmann 1971). Bill's photometer, mounted at the Newtonian focus of the Rockefeller reflector at Boyden Observatory is shown in Figure 2 and described in more detail by Veverka *et al.* (1974).

Our observations of the occultation were successful, and so were those of groups from the University of Texas (Hubbard *et al.* 1972) and Meudon (Combes *et al.* 1971). We obtained simultaneous light curves at three wavelengths, at high time resolution for both the

Figure 2 Bill Liller's portable occultation photometer mounted at the Newtonian focus of the Rockefeller reflector at Boyden Observatory. Dichroic reflectors inside the photometer split the incoming light into three broad wavelength bands, which are then defined by interference filters. Three uncooled photomultipliers, with their pulse counters taped to them, can be seen mounted radially from the sides of the instrument. Light not picked off by the dichroic reflectors proceeds to an eyepiece on the backplate of the photometer and is used for continuous guiding. The size of the eyepiece is the standard 1 1/4 inch, which can be used to establish the scale of the photograph.

immersion (when β Scorpii disappeared behind Jupiter) and emersion (when it reappeared) of the main component of β Scorpii (Liller *et al.* 1974). We also were able to observe the emersion of a fainter star in the system, β Scorpii C. In the process of getting these data, we were all bitten by the "occultation bug," and wanted to do it again. In particular, I was intrigued by the wealth of detail in the data and challenge posed by the observations: all equipment had to be working perfectly at the appointed time. No second chances in this business!

3   LEARNING FROM EXPERIENCE

Following the β Scorpii occultation I finished my degree at Harvard and went to work in Carl Sagan's Laboratory for Planetary Studies at Cornell. Here I could build new photometric instrumentation—my main interest at the time—and work with Joe Veverka and Larry Wasserman on analyzing our β Scorpii data. With such a unique data set, there was a lot for us to learn. Joe and Larry worked on getting temperature, pressure, and number density profiles from the data by numerical inversion of the light curves (Veverka *et al.* 1974). With this we got into a great debate with Andy Young, then at Texas A.&M., over the validity of the inversion technique (Young 1976; Elliot & Veverka 1976). The central issue was whether the light curve spikes were caused by isotropic turbulence in the Jovian atmosphere (in which case the inversion technique would not be valid), or caused by Jovian atmospheric layers (in which case the inversion technique would be valid). More recent work shows the answer lies somewhere in between (French *et al.* 1982 b; Narayan & Hubbard 1988). Other interesting results that we got from our β Scorpii data were (i) a crude measurement of the helium abundance of the Jovian upper atmosphere (Elliot *et al.* 1974); (ii) and the discovery that the data contained "double spikes" that were caused by the individual components of the spectroscopic binary, β Scorpii A_1 and A_2. By combining the double spike separations for the data from both immersion and emersion, we could estimate the masses of the two stars (Elliot *et al.* 1975 a). We also recognized that the narrowest spikes were in fact focused images of the stars and we could get their angular diameters (Elliot *et al.* 1976). We also found that we could correct our data for contamination by the limb light of Jupiter by using the colors to calibrate the amount of light from Jupiter and the star (Elliot *et al.* 1975 c).

On the instrumentation front, my first job was to design and build a *goniometer*—a device for measuring the reflectance of materials under different illumination conditions. These measurements could be used to interpret spacecraft photometry of planetary and satellite surfaces by Joe and his colleagues. Building the goniometer proved good experience for the next project, which was to design and construct a new occultation photometer. The new model would have cooled photomultipliers, so we could work with fainter objects, and the data would be recorded directly onto digital tape, which would be more accurate and eliminate the time-consuming job of digitizing the data from audio tape. The first things we did with the new photometer were the mutual occultations and eclipses of Jupiter's satellites, which occur every six and a half years, when the Earth passes through the equatorial plane of Jupiter. We observed these events from Harvard's Agassiz Station (since renamed Oak Ridge), but had mixed success, since the events occurred low in the sky and the Massachusetts weather could be downright uncooperative (Wasserman *et al.* 1976).

Our next project with this photometer was to observe the lunar occultation of Saturn's satellites from Mauna Kea in March of 1974. We measured the diameters of the visible disks of the five brightest satellites and discovered that Titan's disk is highly limb darkened (Elliot *et al.* 1975 b). Following this work, we planned to observe lunar occultations of Neptune from Australia and South Africa in order to measure the limb darkening of the planet. It was on this expedition that Ted Dunham, who had just come to Cornell as a graduate student, got into occultation work. In Australia we collaborated with Ken Freeman, who had observed the stellar occultation by Neptune in 1968. Bill Liller, being the most skilled observer on our team, was assigned the job of taming the "Grubb," our apellation for the Melbourne reflector, built by Mr. Grubb in 1862 and later moved to Mount Stromlo (King 1979). This telescope did not have a complete main drive gear, but a partial gear, called a sector, that had to be reset every few hours. One of its former speculum mirrors (68% copper and 32% tin) still resided in the dome at the time (see Figure 3).

Unfortunately our Neptune observations were clouded out, both in Australia and at Boyden Observatory in South Africa. After all that effort and no results—should we continue trying to observe occultations?

4   ABOVE THE CLOUDS

If we were to continue with our occultation program, the next event to observe would be an occultation of ε Geminorum by Mars, which would occur in the spring of 1976. The event would be visible in darkness east of the Mississippi. This suggested that Agassiz Station would be the best place for us to observe from, since it had the largest telescope (61-inch) within the zone where the occultation would be observable in darkness, and we needed the large aperture to reduce the scintillation noise, which would be the limiting factor in the quality of our data. Joe thought that we should try to get some results to show NSF and NASA, our sponsors of the unsuccessful Neptune occultations, even though our chances of having clear weather were probably no more than about 30%. However, I had had enough cloudy weather in Australia and South Africa, and didn't think we should be spending the large amount of time preparing for observations that were likely to be unsuccessful. This debate between Joe and myself went on for a while, until we hit upon the idea of observing the occultation from NASA's Kuiper Airborne Observatory (KAO, see Figure 4). In the KAO we could be virtually guaranteed of being above any clouds, since it normally operated at 41,000 ft so that infrared astronomers could be above the water vapor in the Earth's troposphere. Furthermore, the 36-inch telescope in the KAO would be adequate aperture, since the scintillation noise would be greatly reduced at the

Figure 3 Bill Liller—reflected by a speculum mirror of the Melbourne reflector—ponders observing strategy while eating an apple at Mount Stromlo in 1975.

operating altitude, compared with a ground-based site. As an added bonus, the mobility of the KAO meant that we could maneuver to observe the occultation from virtually wherever we chose.

The only problem with using the KAO would be getting NASA's approval to do so; it was not just a matter of making a telephone call (Elliot & Kerr 1987)! Although the initial referee vote was 12-2 against our proposal, through persistent discussion we finally got permission from NASA to use the KAO to observe the ε Geminorum occultation. An important factor in NASA's decision was concern for the Viking spacecraft, which was scheduled to land on Mars on July 4, 1976 (the 200th anniversary of the USA's Declaration of Independence). There had been some concern on the Viking project for the safety of some of the experiments on the atmospheric probe, if the percentage of argon in the Martian atmosphere were high, as some had speculated. With multi-color observations of the ε Geminorum occultation, we could at least put an upper limit on the amount of argon, using the method that we had developed for estimating the helium abundance in Jupiter's atmosphere. To be sure that the technique of obtaining atmospheric temperature profiles for Mars would be valid, I selected a flight path that would make the apparent velocity of the star be perpendicular to the limb of Mars.

Our observations with the KAO gave us excellent data, and an unexpected bonus—the "central flash"—which was visible from the KAO because our flight path put us directly in line with the center of Mars and the star (Elliot *et al.* 1976). In this situation, the Martian atmosphere acts as a giant lens to focus the starlight. Bill Liller was observing the event

Figure 4 The KAO in flight. The open port forward of the wing allows the 36-inch telescope to view the sky with an elevation range between 35° and 75°, while the azimuth range is restricted to ± 2° from the nominal direction. Hence, the direction of flight is used to set the telescope on the azimuth of the desired star.

from Agassiz station, with Costas Papaliolios, whom he had recruited that very afternoon (Liller *et al.* 1978). As luck would have it, the weather was clear at Agassiz after all. It was also clear at the Goddard telescope, where Bob Millis of Lowell Observatory obtained observations (Wasserman *et al.* 1977), at McDonald Observatory (Texas-Arizona Occultation Group 1977), and at several other eastern sites. However, the airborne data were by far the best, because the stable photometric environment greatly reduced the scintillation noise.

We obtained a limit on the argon, but not as stringent as we had hoped because the light curve had many fewer spikes than we had expected from our work with Jupiter. The main result was our discovery of thermal waves in the Martian upper atmosphere, excited by the diurnal heating of the Martian surface, which were confirmed by the Viking entry probe (Elliot *et al.* 1977 c). Furthermore, with data from the large number of sites, we learned the value of multi-station observations in obtaining the planetary diameter and determining the structure around the limb (French & Taylor 1981).

With the demonstration of the validity of occultation results for the Martian atmosphere and the KAO solution to the annoying problem of clouds, the prospects for stellar occultations took a big turn for the better.

5   *GOD MUST BE AN AMERICAN*
The next occultation opportunity occurred less than a year from our observations of ε Geminorum and Mars. Gordon Taylor, of the Royal Greenwich Observatory, had predicted that Uranus would occult SAO 158687 on March 10, 1977, and this event would be visible from areas surrounding the Indian Ocean (Taylor 1973). Again I had applied to use the KAO on the grounds that this would be a unique opportunity to learn about the Uranian upper atmosphere, and we did not want to be thwarted by clouds. The proposal was readily accepted. We would observe in three wavelengths simultaneously, as we had learned was valuable from our β Scorpii experience. Uranus had methane absorption bands in its spectrum (Elliot *et al.* 1977 d), which became more pronounced at longer wavelengths, and the starlight in these bands would be a larger fraction of the total signal. This observation would involve taking the KAO to the southern hemisphere—which also suited infrared observers, who were eager to observe sources in the southern sky.

Preparations for the expedition were proceeding routinely until a plate taken by Otto Franz and Larry Wasserman at Lowell Observatory showed that the position of SAO 158687 was about 1.5 arc seconds further north than given by the SAO Catalog. Hence, the predicted shadow of Uranus (cast in starlight) moved far to the south, with a looming possibility that it would miss the Earth entirely, and no occultation would be visible. This caused much consternation and soul-searching about whether we should proceed with the KAO expedition. The rewards of successful observations were high, so the risk seemed worth taking, since the final prediction from the Lowell group gave 1 to 5 odds that the occultation would be visible, if we could take the KAO well south of Perth, Australia.

Working with me on the KAO observations were Doug Mink and Ted Dunham, who planned to use the data for his Ph.D. thesis. Bob Millis, with whom we were in close contact, would be observing from the Perth Observatory, and other teams from the University of Arizona, India, and France were deployed at various sites in Australia, India, and South Africa. On the KAO, we had to fly several hours before reaching the spot where we could begin our observations. Just after setting up on Uranus and turning on our data recording equipment, we noticed an unexplained dip in the data. This was followed by another, and then another. After eliminating explanations of equipment failure and high, thin cloud, we decided that the dips must be caused by a "belt of satellites" (*sic*) around Uranus. In order to alert other observers of the unexpected brief occultations before it would be too late, I sent the following telegram, via a radio link from the KAO, to Brian Marsden: "Please inform South African observers immediately that secondary occultations of SAO 158687, presumably caused by small bodies in orbit around Uranus, were observed from the Kuiper Airborne Observatory. Suggest that they observe until dawn. Jim Elliot." Just over a half an hour after the last secondary occultation, Uranus occulted the star, which lasted about 25 minutes and relieved the anxiety that had been building for several months.

Although our flight plan had called for an immediate return to Perth following the planetary occultation, I wanted to confirm the secondary occultations by extending our observations of the star as long as possible. Fortunately the KAO flight crew was able to extend our flight leg on Uranus until dawn forced us to stop acquiring data, but this proved long enough to record several more secondary events. After finding that the dips in our data before the planetary occultation matched those after the planetary occultation (see Figure 5) a few days later, we concluded that the explanation must be narrow rings instead of small satellites (Elliot *et al.* 1977 a, b; Elliot & Kerr 1987). Additional data were obtained from Australia, India, and Cape Town South Africa, but other sites had poor weather.

Figure 5 Occultations by the rings of Uranus. The pre-immersion and post-emersion occultations by the rings of Uranus observed with the Kuiper Airborne Observatory (Elliot *et al.* 1977 b) have been plotted on the common scale of distance from the center of Uranus in the ring plane. Occultations corresponding to the nine rings discovered prior to Voyager are easily seen. The low frequency variations in the light curves are due to a variable amount of scattered moonlight on the telescope mirror (after Elliot 1979).

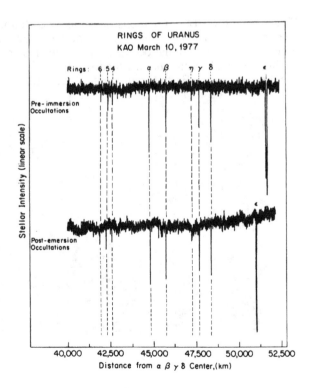

Why were we able to reach the conclusion, in just a few days, that Uranus is encircled by narrow rings? I attribute it to a combination of luck, careful preparation, and the extensive data set that we obtained with the KAO (see Elliot & Kerr 1987). Another explanation, offered by a French astronomer whose observations had been clouded out in South Africa, was related to me later by André Brahic, of the Observatoire de Paris: "God must be an American."

6   MORE OCCULTATIONS ???

The prospects for observing the rings again looked bleak. Bill Sinton, at the University of Hawaii, examined his deepest plates of Uranus (which he had taken from Mauna Kea in search of satellites inside the orbit of Miranda) and found no trace of the rings. Hence, they must reflect very little sunlight. Further observations would have to be done with occultations, at least for the time being, but when would the next occultation occur? As can be seen in Table I, only one occultation by Uranus in 25 years had been identified by Gordon Taylor (see Figure 6), who had been searching for these events since 1952 by comparing planetary ephemerides with star catalogs. At this rate, it would not be until after the Voyager spacecraft could get there (to be launched in late summer 1977) that we could expect to have more observations of the rings.

One way to increase the frequency of predicted occultations by Uranus would be to include more stars in the search, so Brian Marsden asked Arnold Klemola, of Lick Observatory, to take plates of star fields through which Uranus would pass in the near future. From these data they put together a list of twelve stars that would be occulted during the next three years (Klemola and Marsden 1977). However, from Bill Liller's estimates of the magnitudes of these stars from the diameters of their images on the Palomar plates (Liller 1977), I was skeptical that any would be bright enough to yield useful occultation data.

Figure 6 Gordon Taylor demonstrates the "occultation machine" at Herstmonceux in 1975. This mechanical device could be set up with the circumstances of the lunar occultation of a given star and then set in motion to show the path of the moon's shadow across the Earth. Then one could see the zone on the Earth for which the occultation would be observable.

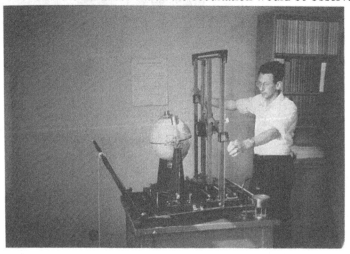

Uranus is so bright, that   a fairly bright star would be needed to provide a signal that would be detectable above the fluctuations in the light from the planet. After calculating the expected occultation depths and signal-to-noise ratios, however, I found that the situation was not totally hopeless and that a 10th magnitude star would suffice (Elliot 1977). In fact, Bob Millis observed the event involving the brightest star in their list from the Canary Islands in December 1977. Although the data were no where near the quality of what we had obtained in March, there definitely were ring occultations in the data, from which we could, at least, learn more about the ring orbits.

7   ENTER INFRARED

Klemola and Marsden's list of Uranian ring occultations proved even more valuable when astronomers at Caltech realized that the methane bands in the Uranian spectrum, which we had used to advantage in the visible region, got even deeper in the infrared. Fortunately, one of these deep methane bands nearly matches the K filter (at 2.2 μm), where the Uranian albedo is so low that the rings are brighter than the planet (Nicholson *et al.* 1978)! Unhampered by the background light from Uranus, they could observe occultations by much fainter stars—especially red ones—than we could use for observations at visible wavelengths, at which all previous observations of ring occultations had been carried out. They planned to observe several of these occultations in the spring of 1978 from Las Campanas in Chile. Eric Persson and Keith Matthews were successful in obtaining observations of the event on April 10, 1978 after getting some marginal results on the April 4 event. Phil Nicholson, under the supervision of Peter Goldreich, analyzed the data as a major part of his Ph.D. thesis at Caltech. The fundamental result here was the demonstration that the unequal radii of the ε ring could be explained by a precessing, eccentric ring. This was surprising, since the common wisdom held that rings should be circular.

Following the Caltech observations, Bill Liller—who had been carrying out several programs at Cerro Tololo—suggested that we could observe Uranian ring occultations from there, in collaboration with Jay Frogel (see Figure 7). We attempted observations of an event in January 1979, but were clouded out. Undaunted, we made further attempts in

Figure 7 Bill Liller and Jay Frogel check out the infrared photometer in the Cassegrain cage of the 4-meter telescope at CTIO (*circa* 1980).

March and August 1980 that proved successful. Data for these events were also obtained by Jay Elias and Keith Matthews from Las Campanas and by Ian Glass from the 1.7 meter telescope at Sutherland. Observers from ESO started regular observations from La Silla. In 1981, David Allen, of the Anglo-Australian Observatory, and I started working at the AAT on events that presented special opportunities. Then Phil went to Mount Stromlo and worked with Terry Jones. We had pieced together a grass roots network that girded the Earth and would be effective while Uranus was near its extreme southern declinations. Fortunately for this project, great improvements were being made in the indium antimonide detectors that we used at 2.2 μm. We could observe from Cerro Tololo, the IRTF in Hawaii, Mount Stromlo, and the AAT—a network that gives widespread longitude coverage, so that virtually any occultation event can be observed when Uranus is not too close to the sun.

Although the diffraction limit was larger because of the longer wavelength of the observations, the signal-to-noise ratio was much better than our original KAO observations. We found that the η ring has two components, the δ ring has a companion ring, and that the α ring has two peaks. We also clearly identified structure in the ε ring by comparison of its occultation profiles from the three Chilean observatories (see Figure 8). Now we could observe the Uranian rings several times a year. The observations were becoming routine, but the new results kept the work exciting.

Figure 8 Occultation profiles for the ε ring emersion. These profiles show the ε ring at a width of 65 km,—about midway between its maximum and minimum widths. The six features marked with dashed lines appear in all profiles and suggest that the ε ring has significant structure below the limit of the resolution imposed by diffraction (after Elliot *et al.* 1983).

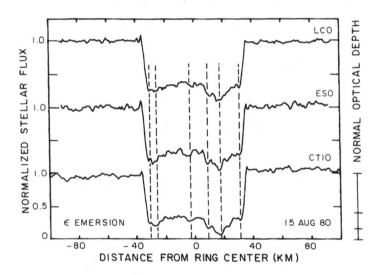

8   CONNECT THE DOTS

As the infrared observations began to flow in regularly, at the rate of 1-2 occultations per year, just how did we use them to learn more about the Uranian rings? The answer to this question is simply that we used all the data to solve a giant "connect-the-dots" puzzle, as illustrated in Figure 9. For each ring occultation we knew one point in space through which the ring orbit passed, and by combining these points with the constraints that the rings were in orbit about Uranus, we could solve for the orbital parameters for each ring. The first assumption was that the rings followed circular orbits, but with this assumption we could see a problem from the start: the ε ring could not be modeled as a circle—it must be eccentric, inclined relative to the orbits of the other rings, or be a discontinuous ring (Elliot *et al.* 1978). This problem was resolved by Phil Nicholson and his colleagues at Caltech, who combined the data from the first three observed occultations and found that the ε ring is eccentric (Nicholson *et al.* 1978). Not only that, but its variable width was correlated with its orbit: it was only about 20 km wide at periapse, but this width increased to nearly 100 km at apoapse. Since the orbit of each particle in the ring would precess at a rate that depended on its distance from the planet, the regularity of the ε ring width meant that somehow the precession of all the orbits was "locked" together.

As the data continued to be added, the sophistication of our orbit model increased, since we were able to keep combining the data from all the events into a single model. We began with treating each ring as a precessing ellipse (Elliot *et al.* 1981 a), and then showed that several of the rings had inclined orbits (French *et al.* 1982 a). Inclinations of ring orbits had been dismissed as impossible on theoretical grounds, in the past. The continual improvement in our model is shown in Figure 10, where the precision versus year is plotted. The rings that had high residuals from our model were scrutinized carefully, and

Figure 9 Observed occultation points in the sky plane at Uranus. The observed points are marked by open circles, pre-immersion points lie to the right and post-emersion points to the left. The dashed lines indicate the tracks of the observatories in the sky plane. The rings are indicated by dotted lines, except for the observed immersion and emersion segments, which are solid lines. The radius of Uranus indicated is the radius of the "half-light" level, corresponding to a number density of 6×10^{13} cm^{-3}. The "north pole" in the figure has become the south pole, according to a more recent definition by the IAU (after Elliot *et al.* 1978).

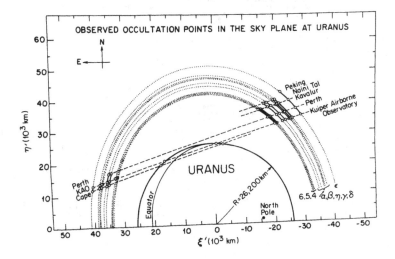

we have found perturbations on the γ and δ rings that follow a regular pattern (French *et al.* 1986).

The Uranian rings inspired several theoretical attempts to explain how such narrow rings could remain stable without spreading apart. The most widely accepted explanation is the "shepherd satellite" model of Goldreich and Tremaine (1979). In their picture, each narrow ring is maintained by two small satellites, one orbiting inside and the other orbiting outside the ring. Angular momentum is transferred from the inner satellite to the ring and from the ring to the outer satellite, which results in the ring being kept narrow because its particles are being pushed out by the inner satellite and in by the outer one.

In addition to learning more about the rings themselves from modeling their orbits, we have used the rings as test particles to probe the gravity field of Uranus. Since the ring precessions are caused by deviations of the gravity field from spherical symmetry— primarily due to the oblateness of the planet—our orbit model gives precise values for the gravity harmonics J_2 and J_4 for Uranus. These values constrain interior models, so that we know that Uranus has a small rocky core, surrounded by more volatile materials. It is ironic that what looked like an occultation that had questionable chances for success ultimately resulted in learning much more about Uranus than we would have dared to imagine—even in our most optimistic moments.

9   *HAVE TELESCOPE, WILL TRAVEL*

Another main thrust for the stellar occultation technique has been the determination of diameters and elliptical shapes of asteroids. These quantities are calculated by fitting an ellipse through the end-points of several occultation chords observed from sites

Figure 10 Development of ring orbit models. As the model has been made more realistic by including more parameters, the rms error per degree of freedom has decreased. However, the rms error of the present model is still substantially larger than the timing errors in the data, indicating that the model is not yet complete (after Elliot & Nicholson 1984).

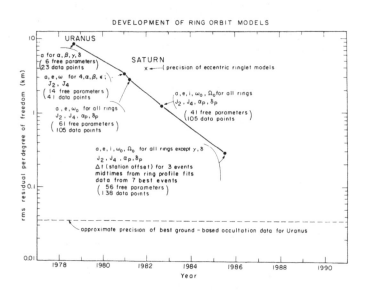

that are (ideally) uniformly spaced across the asteroid's shadow, perpendicular to its track on the Earth. From an accurate asteroid diameter, we can learn the albedo of the asteroid, and, if the asteroid's mass is known, we can also learn its density—quantities that limit the possibilities for the composition of the asteroid. The challenges for observing stellar occultations by asteroids arises from their small sizes (the mean diameter of the largest ten asteroids is just over 400 km; Millis & Elliot 1979), which means that (i) the occultation prediction must be reliable to better than a few tenths of an arc second, and (ii) several observations of the occultation must be made within the width of the asteroid's shadow. The first asteroid occultation observations were visual. The observers were alerted to these events by Gordon Taylor, who compared the SAO catalog with asteroid ephemerides. However, for these only one or two chords were observed, and the observations were made by visual observers, which proved to be not reliable enough (Millis & Elliot 1979; Taylor & Dunham 1978). A notable exception to the sparse chord problem was the occultation of κ Geminorum by Eros, for which 8 chords were obtained by observers equipped with binoculars and stop watches (O'Leary et al. 1976).

The first major effort to determine a reliable asteroid diameter by photoelectric observations of an asteroid occultation occurred for the Pallas occultation of SAO 85009 in 1978. Special astrometric plates were taken to refine the prediction as Pallas approached the star. In all, observations were attempted from 22 sites that were near the predicted path of Pallas's shadow, that went nearly east-west across the United States (Wasserman et al. 1979). Four of these were mobile observatories—the KAO, from which we observed the northern most chord; and the other three were expeditions launched by Bob Millis of Lowell Observatory, Harold Reitsema, of the University of Arizona, and Richard Baron, now at MIT. Of the total sites, 7 were successful in obtaining photoelectric observations of the occultation. (One of the northern sites, which was unfortunately north of the shadow, was at 77 Snake Hill Road Belmont Mass and manned by Bill Liller.) The result of these efforts was a mean diameter of 538 ± 12 km, which implied a mean geometric albedo of 0.103 ± 0.005 at V and a mean density of 2.8 ± 0.5 gm cm^{-3}. These parameters differ markedly from those of Jupiter's Galilean satellites, but are similar to those of the Earth's moon.

Our Pallas experience showed the value of portable telescopes for asteroid occultation work. Shortly thereafter, major efforts to construct sets of portable telescopes, photometers, and data recording equipment occurred at Lowell Observatory, MIT, the University of Arizona, and other institutions. The standard system used in the field is a 14-inch reflecting telescope, with a lightweight photometer. The timing reference is a crystal clock that is set by WWV radio signals shortly before the observations commence. Chasing down asteroid occultations has proven to be "best in the west," where the weather is consistently more favorable and it is easier to go large distances by vehicles after the predicted path of the asteroid shadow has been determined by photographic astrometry shortly before the event. Over the years, it has been possible to field 5-10 portable telescopes for high quality asteroid occultations, and several reliable diameters have been established by stellar occultations, most under the leadership of Bob Millis at Lowell (Millis & Dunham 1988). Many of these are not SAO stars, but have been found by searching plates for events by Larry Wasserman and his colleagues at Lowell Observatory (Wasserman et al. 1987).

An interesting possibility that was suggested by some early occultation observations is that some asteroids are not simple bodies, but multiple. This conclusion is based on dips in the signal outside the main asteroid occultation. In many (if not all) cases, the dips in signal have explanations other than being caused by multiple bodies comprising what appears to be a single asteroid at low resolution. Other evidence for multiplicity has come from speckle interferometry and the shapes of asteroid light curves. However, at least from the occultation observations, there has been no unambiguous data, so that most are not convinced that the suspected satellites of asteroids really exist. This issue could be resolved positively by future convincing observations, while some of the reported cases could be

resolved negatively by the high resolution imagery that should become possible with the
Hubble Space Telescope.

10   *VOYAGER'S ADVENTURES*

Having seen the advantages for stellar occultation observations provided by
the KAO and portable telescopes, we should not be surprised to learn that observing
occultations from another mobile observatory—the Voyager 2 spacecraft—would produce
some unique and exciting results. The Voyager instruments suitable for stellar occultation
observations are (i) the PPS (photo-polarimeter subsystem) and (ii) the UVS (ultraviolet
spectrometer). The observation procedure is the same as observing an occultation from the
Earth, namely one records the intensity of the star at high time resolution as it disappears
behind a ring system or planetary atmosphere. From a technical perspective, stellar
occultation observations by a flyby spacecraft have the disadvantage that the light collecting
ability of the telescope is much smaller, increasing the relative amount of photon noise;
however this disadvantage is offset by the advantages: (i) a large number of stars are
occulted from the vantage of a flyby spacecraft, because of the proximity of its target; hence
bright occultation candidate stars are available; (ii) the Fresnel diffraction limit on spatial
resolution is typically a few tens of meters, again because of the proximity; and (iii) the
occultation can be observed at ultraviolet wavelengths, where reflected sunlight is less,
relative to available occultation candidates with early spectra. This gives a larger stellar
flux, relative to the background light of the occulting body. A difference for observing an
atmospheric occultation from a spacecraft is that one probes the extinction by the
atmospheric constituents, rather than the refraction, for two reasons. First, the optical
depths of the atmospheric gases are generally greater in the uv, and more importantly
because the observations are made relatively close to the occulting body, which provides
only a short lever arm for refraction.

At Saturn, Voyager 2 observed the occultation of δ Scorpii by the rings and achieved a
spatial resolution of 100 meters, about 30,000 times better than the best ground-based
images and 100 times better than the Voyager imaging observations (Lane *et al.* 1982;
Sandel *et al.* 1982). With this spatial resolution they could unmistakably identify the
patterns of spiral density waves in the rings, even as the wavelength decreased in the more
tightly wound portions. These data also revealed fine structure in the enigmatic F ring,
showing that particles are concentrated at the edges of the ringlet. Other observations at
Saturn by Voyager that were related to the scientific issues addressed by stellar occultations
were the discovery of the F ring shepherd satellites by Voyager 1 and the occultation of the
spacecraft radio signals recorded by the RSS (radio science) team as the spacecraft passed
behind the rings and planet from the perspective of the radio antennas on Earth. Although
the Fresnel diffraction limit for the much longer radio wavelengths is several kilometers, a
crucial element to the radio observations is that the phase of the wave, as well as its
intensity are observed. Hence the data can be deconvolved to a spatial resolution even
better than that achievable by the PPS data. (Tyler *et al.* 1983; Marouf *et al.* 1986). Finally,
the large span of wavelengths, from ultraviolet to radio, covered by the Voyager
instruments enable one to learn about the sizes of particles that comprise the rings (Marouf
et al. 1983).

The Uranus encounter by Voyager 2 occurred in January 1986, and ground-based
observers had amassed much information to aid the encounter science investigations (Elliot
1985). Two stellar occultations by the rings were observed by Voyager 2: (i) one of σ
Sagittarii by the δ and ϵ rings that had by far the best signal-to-noise ratio, and (ii) one of β
Persei by the entire ring system (Lane *et al.* 1986; Holberg *et al.* 1987). A notable result
from the σ Sgr occultation was the additional fine structure observed in the two rings
probed. Combining the UVS and PPS occultation data for 1986U1R, a new ring
discovered by the Voyager imaging team, revealed that this ring is quite different from the
others. It has an optical depth that is a strong function of wavelength, indicating that it

consists of submicron particles, while the other rings are composed mostly of particles larger than a few centimeters. The Voyager radio occultation observations can be deconvolved to the best spatial resolution achieved by the spacecraft (Gresh *et al.* 1988). Related to the ring system are the ten new Uranian satellites discovered by Voyager, whose orbits are shown in Figure 11 along with those of the rings. In this figure, the orbits appear as a uniform progression, with the orbits nearer the planet being those of rings and the outer ones being the satellite orbits. In the region of the ε ring, both satellite and ring orbits exist.

At the time of this writing, Voyager is headed for its last hurrah, an encounter with Neptune on August 29, 1989. There it may answer many questions posed by the Earth-based occultation observations described in the next section.

11 *NEPTUNE'S "ARCS"*

Although Neptune had been an early target for atmospheric investigation with a stellar occultation (Freeman & Lyngå 1970), interest in searching for a ring system developed only after the Uranian ring discovery and a list of potential occultations was generated from photographic plates (Klemola *et al.* 1978; Mink *et al.* 1981). Two major efforts were mounted to observe stars as they probed the region of potential rings around Neptune. One in 1981, involved simultaneous observations from telescopes in Australia at Siding Spring and Mount Stromlo, and from several telescopes at Mauna Kea (where Bill Liller observed this event with the 2.2-meter telescope). These data were combined with those obtained by Jay Elias with the 4-meter telescope at Cerro Tololo for another occultation two weeks later. No evidence for rings was found (Elliot *et al.* 1981 b; Hubbard *et al.* 1985). Since this search probed for continuous rings only within about 6000 km of the top of Neptune's atmosphere, another major effort that would probe closer to Neptune was mounted in 1983. This occurred in the same part of the world, and in

Figure 11 Rings and inner satellites of Uranus. The orbits of the Uranian satellites discovered by Voyager, with their provisional designations, are shown with the orbits of the main rings. The synchronous orbit and the Roche radii for fluid bodies of densities of 1 and 2 gm cm^{-3} have been drawn for comparison. (after Elliot & Kerr 1987).

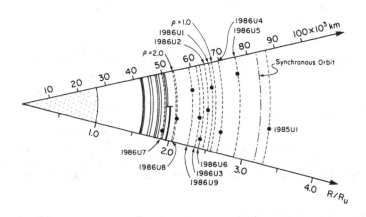

addition to the telescopes used in the previous search, the KAO was brought into service, since its flight path could be positioned to probe for rings in the equatorial plane of Neptune, down to the very top of Neptune's atmosphere. Again, no rings; an upper limit of 0.016 was placed on the optical depth of continuous equatorial rings (Elliot et al. 1985).

The next opportunity would occur in July of 1984. At this point I had had enough of the Neptune ring search, considering that two extensive efforts had turned up no evidence for rings. In fact, I had written "If Neptune has rings, they almost certainly will not be discovered from the ground" (Elliot & Kerr 1987). Fortunately, however, observations were arranged by André Brahic to be carried out at ESO and by Bill Hubbard to be carried out at Cerro Tololo. The ESO observers reported a dip in their data that they thought might be a small satellite, "although it cannot be completely excluded that the occultation was caused by part of an irregular ring," (Manfroid et al. 1984; Häfner & Manfroid 1984). However, no one in the community paid much attention to these reports, since it was not unusual to hear about "unexplained dips" that almost invariably had turned out to have been caused by instrumental problems or unnoticed clouds. So, it was not for several months until Bill Hubbard examined the CTIO data and found a nearly identical dip at nearly the identical time (Hubbard et al. 1984)! Publication of an analysis of these results was considerably delayed due to a protracted battle among the organizers and observers about how the results should be published (Maddox 1985). Since previous searches had ruled out complete rings, the conclusion from these data was that Neptune has at least one incomplete ring, dubbed an "arc" (Hubbard et al. 1986). Theoreticians were quick to devise several mechanisms by which such an incomplete ring could have a stable existence (Lissauer 1985; Goldreich et al. 1986; Lin et al. 1987).

Following this discovery, attempts have been made to observe the arcs again. There are 6 plausible observations of arcs out of 21 occultations observed (Nicholson et al. 1987), so the accumulation of information has been slow. Arcs of at least three, possibly four radii are required to account for the observations, although moderately eccentric and/or inclined orbits would reduce the number of independent arcs observed to date (Nicholson et al. 1987). If we could image the arcs, their orbits could be established quickly. So far, however, imaging has not been successful, and our understanding of the arcs may not develop further until the Voyager encounter of the Neptune system in August 1989.

12   WHAT NEXT?

As we have seen for the past, new results from stellar occultations depend on (i) predictions of occultations by interesting bodies, (ii) mobile observatories, (iii) high quality data, and (iv) innovative methods of data analysis. In the future, we can expect more exciting discoveries from stellar occultations, if we can find ways for improving our techniques in these areas.

For example, consider occultation predictions. At first, we all depended on the predictions for catalog stars generated by Gordon Taylor, and further observations of occultations by asteroids and the outer planets did not occur until more events were identified with the introduction of the photographic technique. Right now we are limited by the accuracy of the prediction that we can attain with the photographic method—about 0.2 arcsec. This error is several times larger than the radii of some interesting bodies to probe with occultations, such as Pluto and Neptune's satellite Triton. Through observation of occultations by these bodies we could answer such fundamental questions as, "does Pluto have an atmosphere?" We know that stars are occulted by Pluto that are bright enough to produce high-quality data (Bosh et al. 1986), but we don't know where to go to observe the events!

One hope for improving the accuracy of occultation predictions is a new method for astrometry, known as "CCD strip scanning." Here we lock the telescope and let the sky drift by, similar to the operation of a transit circle. The rows of the CCD are clocked at the same rate as the stars move across the chip, so one generates a long, narrow CCD frame on which the stellar positions should have precise relative positions. If these strip scans are recorded as the occulting body approaches its target, we should be able to generate a quite accurate prediction. A test of this method was carried out during the spring of 1988, in an attempt to generate an accurate prediction for the occultation by Pluto on 1988 June 9. As this article goes to press, we note that the Pluto occultation has been successfully observed. The slow disappearance of the starlight showed that Pluto does indeed have an atmosphere (Elliot & Dunham 1988; Millis 1988; Blow 1988). The final prediction done with photographic plates proved more reliable than that made on the basis of CCD data, and more work will be needed to understand the errors involved with the CCD strip scans.

The CCD should also prove valuable in obtaining better data in situations involving large gradients in the background light, such as stellar occultations by Saturn's rings. Using a high-speed CCD, such as that described by Dunham *et al.* (1985), and modeling the background of the rings, one could achieve much better photometry than with a conventional aperture photometer. So far, no occultations by Saturn's rings have been observed from the Earth with good photometric instrumentation, and, as can be seen in Figure 10, the accuracy of kinematical models for Saturn's rings lags far behind those for Uranus. If we can succeed with CCD techniques for regular observations of occultations by Saturn's rings, we should learn a lot more about this complex ring system.

Occultation observations will benefit from several of the new telescopes planned and under construction. The Hubble Space Telescope, not being plagued with the seeing and scintillation problems caused by the Earth's atmosphere, will be ideal for observing occultations by Saturn's rings. The amount of background light from the rings will be greatly reduced by using a small focal plane aperture. The HST will have the disadvantage, however, of its view being blocked by the Earth about half the time, so that the occultation data it records will be periodically interrupted. The large aperture telescopes, such as the Keck Telescope and others, will reduce the scintillation and photon noise, permitting observations of more events with better signal-to-noise ratio. This should be helpful for Uranian ring studies, since fainter stars generally have smaller angular diameters, and the stellar diameter of the occulted star has limited the spatial resolution for most of our past observations.

Considering mobile observatories, we can anticipate that the 3-meter airborne telescope (SOFIA) currently under consideration by NASA, would combine the advantages of large aperture and mobility. Plans for the Cassini mission, a spacecraft that would orbit the Saturn system for more than a year, now include an occultation photometer. This instrument would achieve ultra-high spatial resolution and longer term coverage than a flyby spacecraft.

Finally, the development of the stellar occultation technique might have been considerably delayed if it weren't for Bill Liller's contributions to the early work described in this review. In particular, his multi-channel, portable photometer that he assembled for the β Scorpii occultation was essential to obtaining the fascinating data set for this event, which inspired many of the new approaches for learning more from occultation data than had been anticipated. Also, his continuing collaboration in the observations was a source of inspiration for many of us—for me in particular.

Thanks, Bill.

REFERENCES

Baum, W.A. & Code, A.D. (1953). A photometric observation of the occultation of σ Arietis by Jupiter. *Astron. J.*, **58**, 108-112.

Blow, G.L. (1988). Occultation by Pluto. *IAU Circ.* No. 4611.

Bosh, A.S., Elliot, J.L., Kruse, S.E., Baron, R.L., Dunham, E.W. & French, L.M. (1986). Signal-to-noise ratios for possible stellar occultations by Pluto. *Icarus*, **66**, 556-560.

Brinkmann, R.T. (1971). Occultation by Jupiter. *Nature*, **230**, 515-516.

Combes, M., Lecacheux, J. & Vapillon, L. (1971). First results of the occultation of β Sco by Jupiter. *Astron & Astrophys.*, **15**, 235-238.

Dunham, E.W., Baron, R.L., Elliot, J.L., Vallerga, J.V., Doty, J.P. & Ricker, G.R. (1985). A high-speed, dual-CCD imaging photometer. *P.A.S.P.*, **97**, 1196-1204.

Elliot, J.L. (1977). Signal-to-noise ratios for occultations by the rings of Uranus 1977-1980. *Astron J.*, **82**, 1036-1038.

Elliot, J.L. (1979). Stellar occultation studies of the solar system. In *Annual Reviews of Astronomy and Astrophysics*, **17**, 445-75.

Elliot, J.L. (1985). Uranus: The view from Earth. *Sky and Telescope*, **70**, 415-419.

Elliot, J.L., Baron. R.L., Dunham, E.W., French, R.G., Meech, K.J., Mink, D.J., Allen, D.A., Ashley, M.C.B., Freeman, K.C., Erickson, E.F., Goguen, J. & Hammel, H.B. (1985). The 1983 June 15 occultation by Neptune: Limits on a possible ring system. *Astron J.*, **90**, 2615-2623.

Elliot, J.L., & Dunham, E. W. (1988). Occultation by Pluto. *IAU Circ.* No. 4611.

Elliot, J.L., Dunham, E.W. & Church, C. (1976). A unique airborne observation. *Sky and Telescope*, **52**, 23-25.

Elliot, J.L., Dunham, E.W. & Millis, R.L. (1977 a). Discovering the rings of Uranus. *Sky and Telescope*, **53**, 412-416, 430.

Elliot, J.L., Dunham, E.W. & Mink, D.J. (1977 b). The rings of Uranus. *Nature*, **267**, 328-330.

Elliot, J.L., Dunham, E.W., Wasserman, L.H., Millis, R.L. & Churms, J. (1978). The radii of Uranian rings α, β, γ, δ, ε, η, 4, 5, and 6 from their occultations of SAO 158687. *Astron J.*, **83**, 980-992.

Elliot, J.L., Elias, J.H., French, R.G., Frogel, J.A., Liller, W., Matthews, K., Meech, K.J., Mink, D.J., Nicholson, P.D. & Sicardy, B. (1983). The rings of Uranus: Occultation profiles from three observatories. *Icarus*, **56**, 202-208.

Elliot, J.L., French, R.G., Dunham, E.W., Gierasch, P.J., Veverka, J., Church, C. & Sagan, C. (1977 c). Occultation of ε Geminorum by Mars. II. The structure and extinction of the Martian upper atmosphere. *Astrophys. J.*, **217**, 661-679.

Elliot, J.L., French, R.G., Frogel, J.A., Elias, J.H., Mink, D.J. & Liller, W. (1981 a). Orbits of nine Uranian rings. *Astron J.*, **86**, 444-455.

Elliot, J. & Kerr, R. (1987). *Rings: Discoveries from Galileo to Voyager.* (first paperback edition) Cambridge: MIT Press.

Elliot, J.L., Mink, D.J.,Elias, J.H., Baron, R.L., Dunham, E., Pingree, J.E., French, R.G., Liller, W., Nicholson, P.D., Jones, T.J., & Franz, O.G.(1981 b). No evidence of rings around Neptune. *Nature*, **294**, 526-529.

Elliot, J.L. & Nicholson, P.D. (1984). The rings of Uranus. In *Planetary Rings*, ed. R. Greenberg and A. Brahic, pp. 25-72. Tucson: University of Arizona Press.

Elliot, J.L., Rages, K. & Veverka, J. (1975 a). Occultation of β Scorpii by Jupiter. VI. The masses of β Scorpii A_1 and A_2. *Astrophys. J. (Letters)*, **197**, L123-126.

Elliot, J.L., Rages, K. & Veverka, J. (1976). Occultation of β Scorpii by Jupiter. VII. The angular diameters of β Scorpii A_1 and A_2. *Astrophys. J.*, **207**, 994-1001.

Elliot, J.L. Veverka, J. & Goguen, J.(1975 b).Lunar occultation of Saturn. I. The diameters of Tethys, Dione, Rhea, Titan, and Iapetus. *Icarus*, **26**, 387-407.

Elliot, J.L. & Veverka, J. (1976).Stellar occultation spikes as probes of atmospheric structure and composition. *Icarus*, **27**, 359-386.

Elliot, J.L., Veverka, J. & Millis, R.L. (1977 d). Uranus.occults SAO 158687 *Nature*, **265**, 609-611.

Elliot, J., Wasserman, L.H., Veverka, J., Sagan, C. & Liller, W., (1974). The occultation of β Scorpii by Jupiter. II. The hydrogen-helium abundance in the Jovian atmosphere *Astrophys J.*, **190**, 719-729.

Elliot, J., Wasserman, L.H., Veverka, J., Sagan, C. & Liller, W., (1975 c). Occultation of β Scorpii by Jupiter. V. The emersion of β Scorpii C. *Astron. J.*, **80**, 323-332.

Freeman, K.C. & Lyngå, G. (1970). Data for Neptune from occultation observations. *Astrophys. J.*, **160**, 767-780.

French, R.G., Elliot, J.L. & Allen, D.A. (1982 a). Inclinations of the Uranian rings. *Nature*, **298**, 827-829.

French, R.G., Elliot, J.L. & Levine, S. (1986). Structure of the Uranian rings. II. Ring orbits and widths. *Icarus*, **67**, 134-163.

French, R.G., Elliot, J.L., Sicardy, B. Nicholson, P. & Matthews, K. (1982 b). The upper atmosphere of Uranus: A critical test of isotropic turbulence models. *Icarus*, **51**, 491-508.

French, R.G. & Taylor, G.E. (1981). Occultation of ϵ Geminorum by Mars. IV. Oblateness of the Martian upper atmosphere. *Icarus*, **45**, 577-583.

Goldreich, P. & Tremaine, S. (1979). Towards a theory for the Uranian rings. *Nature*, **277**, 97-99.

Goldreich, P., Tremaine, S. & Borderies, N. (1986). Towards a theory for Neptune's arc rings. *Astron. J.*, **92**, 490-494.

Gresh, D.L., Marouf, E.A., Tyler, G.L., Rosen, P.A. & Simpson, R.A. (1988). Voyager radio occultation by Uranus' rings. I. Observational results. *Icarus* (submitted).

Häfner, R. & Manfroid, J. (1984). 1984 N1. *IAU Circ.* No. 3968.

Holberg, J.B., Nicholson, P.D., French, R.G. & Elliot, J.L. (1987). Stellar occultation probes of the Uranian rings at 0.1 and 2.2 µm: A comparison of Voyager and Earth-based results. *Astron. J.*, **94**, 178-188.

Hubbard, W.B., Frecker, J.E., Gehrels, J.-A., Gehrels, T., Hunten, D.M., Lebofsky, L.A., Smith, B.A., Tholen, D.J., Vilas, F., Zellner, B., Avey, H.P., Mottram, K., Murphy, T., Varnes, B., Carter, B., Nielsen,, A., Page, A.A., Fu, H.H., Wu, H.H., Kennedy, H. D., Waterworth, M.D., and Reitsema, H.J. (1985). Results from observations of the 15 June 1983 occultation by the Neptune system. *Astron. J.*, **90**, 655-667.

Hubbard, W.B., Brahic, A., Sicardy, B., Elicer, L.–R., Roques, F. & Vilas, F. (1986). Occultation detection of a neptunian ring-like arc. *Nature*, **319**, 636-640.

Hubbard, W.B., Nather, R.E., Evans, D.S., Tull, R.G., Wells, D.C., van Citters, G.W., Warner, B. & Vanden Bout, P. (1972). The occultation of Beta Scorpii by Jupiter and Io. *Astron J.*, **77**, 41-59.

Hubbard, W.B., Vilas, F. & Elicer, L.–R. (1984). Probable ring of Neptune (1984 N1, 1981N1). *IAU Circ.* No. 4022.

King, H.C. (1979). *The History of the Telescope.* (Dover edition) New York: Dover.

Klemola, A.R., Liller, W., Marsden, B.G. & Elliot, J.L. (1978). Predicted occultations by Neptune, 1978-1980. *Astron. J.*, **83**, 205-207.

Klemola, A.R.& Marsden, B.G. (1977). Predicted occultations by the rings of Uranus, 1977-1980. *Astron. J.*, **82**, 849.

Lane, A.L., Hord, C.W., West, R.A., Esposito, L.W., Coffeen, D.L., Sato, M., Simmons, K.E., Pomphrey, R.B. & Morris, R.B. (1982). Photopolarimetry from Voyager 2: Preliminary results on Saturn, Titan, and the rings. *Science*, **215**, 537-543.

Lane, A.L., Hord, C.W., West, R.A., Esposito, L.W., Simmons, K.E., Nelson, R.M., Wallis, B.D., Buratti, B.J., Horn, L.J., Graps, A.L. & Pryor, W.R. (1986). Photopolarimetry from Voyager 2: Initial results from the Uranian atmosphere, satellites, and rings. *Science*, **233**, 65-70.

Liller, W. (1977). Colors and magnitudes of stars occulted by the rings of Uranus, 1977-1980. *Astron. J.*, **82**, 929.

Liller, W., Elliot, J.L., Veverka, J., Wasserman, L. & Sagan, C. (1974). The Occultation of Beta Scorpii by Jupiter. III. Simultaneous high time-resolution records at three wavelengths. *Icarus, ***22**, 82-104.

Liller, W., Papaliolios, C., French, R.G., Elliot, J.L. & Church, C. (1978). The Occultation of ε Gem by Mars as observed from Agassiz Station. *Icarus,* **35**, 395-399.

Lin, D.N.C., Papaloizou, J.C.B. & Ruden, S.P. (1987). On the confinement of planetary arcs. *Mon. Not. R. Astr. Soc.*, **227**, 75-95.

Lissauer, J.J. (1985). Shepherding model for Neptune's arc ring. *Nature*, **318**, 544-545.

Maddox, J. (1985). Whose rings of Neptune? *Nature*, **318**, 505.

Manfroid, J., Gutierrez, F., Häfner, R. & Vega, R. (1984). Appulse of SAO 186001 to Neptune. *IAU Circ.* No. 3962.

Marouf, E.A., Tyler, G.L. & Rosen, P.A. (1986). Profiling Saturn's rings by radio occultation. *Icarus*, **68**, 120-166.

Marouf, E.A., Tyler, G.L., Zebker, H.A., Simpson, R.A. & Eshleman, V.R. (1983). Particle size distributions in Saturn's rings from Voyager 1 radio occultation. *Icarus*, **54**, 189-211.

Millis, R.L. (1988). Occultation by Pluto. *IAU Circ.* No. 4611.

Millis, R.L. & Dunham, D. W. (1988) Precise Measurement of Asteroid Sizes and Shapes from Occultations. In *Asteroids II* ed. R. P. Binzel, T. Gehrels, and M. Matthews, Tucson: University of Arizona Press (in press).

Millis, R.L. & Elliot, J.L. (1979). Direct determination of asteroid diameters from occultation observations. In *Asteroids*, ed. T. Gehrels, pp. 98-118. Tucson: University of Arizona Press.

Mink, D.J., Klemola & Elliot, J.L. (1981). Predicted occultations by Neptune: 1981-1984. *Astron. J.*, **86**, 135-137.

Narayan, R. & Hubbard, W.B. (1988). Theory of anisotropic refractive scintillation— application to stellar occultations by Neptune. *Astrophys. J.* (in press).

Nicholson, P.D., Cooke, M.L., Matthews, K., Elias, J. & Buie, M.W. (1987). Models for Neptune's arc-rings. *Bull. Am. Astr. Soc.*, **19**, 888.

Nicholson, P.D., Persson, S.E., Matthews, K., Goldreich, P.D. & Neugebauer, G. (1978). The rings of Uranus: Results of the 10 April 1978 occultation. *Astron. J.*, **83**, 1240-1248.

O'Leary, B., Marsden, B.G., Dragon, R., Hauser, E., McGrath, M., Backus, P. & Robkoff, H. (1976). The occultation of κ Geminorum by Eros. *Icarus*, **28**, 133-146.

Sandel, B.R., Shemansky, D.E., Broadfoot, A.L., Holberg, J.B., Smith, G.R., McConnell, J.C., Strobel, D.F., Atreya, S.K., Donahue, T.M., Moos, H.W., Hunten, D.M., Pomphrey, R.B. & Linick, S. (1982). Extreme ultraviolet observations from the Voyager 2 encounter with Saturn. *Science*, **215**, 548-553.

Taylor, G.E. (1973). An occultation by Uranus. *J. Brit. Astron. Assoc.*, **83**, 352.

Taylor, G.E. & Dunham D.W. (1978). The size of minor planet 6 Hebe. *Icarus*, **34**, 89-92.

Texas-Arizona Occultation Group (1977). The occultation of Epsilon Geminorum by Mars: Analysis of McDonald data. *Astrophys. J.*, **214**, 934-945.

Tyler, G.L., Marouf, E.A., Zebker, H.A., Simpson, R.A. & Eshleman, V.R. (1983). The microwave opacity of Saturn's rings at wavelengths of 3.6 and 13 cm from Voyager 1 radio occultation. *Icarus,* **54**, 160-188.

Veverka, J., Wasserman, L.H., Elliot, J., Sagan, C. & Liller, W., (1974). The occultation of β Scorpii by Jupiter. I. The structure of the Jovian upper atmosphere. *Astron. J.*, **79**, 73-84.

Wasserman, L.H., Bowell, E. & Millis, R.L.(1987). Occultations of stars by solar system objects. VII. Occultations of catalog stars by asteroids in 1988 and 1989. *Astron. J.*, **94**, 1364-1372.

Wasserman, L.H., Elliot, J.L., Veverka, J. & Liller, W. (1976). Galilean satellites: Observations of mutual occultations and eclipses in 1973. *Icarus*, **27**, 91-107.

Wasserman, L.H., Millis, R.L.Franz, O.G., Bowell, E., White, N.M., Giclas, H.L., Martin, L.J., Elliot, J.L., Dunham, E., Mink, D., Baron, R., Honeycutt, R.K., Henden, A.A., Kephart, J.E., A'Hearn, M.F. & Reitsema, H.J. (1979). The diameter of Pallas from its occultation of SAO85009. *Astron. J.*, **84**, 259-268.

Wasserman, L.H., Millis, R.L. & Willamon, R.M. (1977). Analysis of the occultation of ε Geminorum by Mars. *Astron. J.*, **82**, 506-510.

Young, A.T. (1976). Scintillations during occultations by planets. I. An approximate theory. *Icarus*, **27**, 335-357.

QUESTIONS AND COMMENTS

Papaliolios: This is just a historical note, a testimonial on Bill's ability to mount complex observing programs on short notice. Bill called me one afternoon at about 4 pm, and asked if I would like to do an "occultation" with him. I said, sure - when? He replied how about tonight! I called home, telling my wife not to expect me for dinner, and we obtained that night some of the best data on the occultation of ε Geminorum by Mars. This was done with the photometer that Paul Horowitz and I had been using at that period to make our pulsar observations. In addition, I'd like to verify that Jim Elliot had tied up the world's telescopes to look at the rings of Uranus on April 1 (2 months ago). I was at CTIO to make observations on the Supernova 1987A, and had to sit by while the Uranus observations took place.

Elliot: The "April Fool's Day" occultation by Uranus and its rings was particularly interesting because it occurred as Uranus was at the end of its retrograde loop. Instead of the usual 2 hours between immersion and emersion for the ε ring, the time between these events was 4 days! Hence the relative velocity of the star and rings was about 1 km sec[-1] for the ring occultation events, and, with low velocity, we were able to probe the longitudinal structure of the rings. Under these special circumstances, we found "kinks" on the ε ring profile edges, which are probably indicative of low-amplitude waves on the ring edges that are excited by one or more small, nearby satellites. We had a similar conflict with supernova observations at Siding Spring (with one of your speckle competitors), but the director was able to work out a compromise so that we all got the data we wanted. Fortunately Uranus and the supernova were on opposite sides of the sky.

Liller: Carl Sagan was also involved in the early planetary occultation work, and in fact was project director of the Jupiter expedition to South Africa. And of course, he hired Jim and Joe Veverka at Cornell. For the record, I should like to add that after Joe and I had made an unsuccessful attempt to identify the pulsar CP1919, a short note appeared on page 56, as I remember, in the N.Y. Times. Johnny Carson happened to see it and being an amateur astronomer, invited me to appear on his show. One person that saw it was Carl Sagan, and according to what Carl told me, it was then and there that he got the idea to begin his TV career. The rest is, as they say, history.

Elliot: Yes, Carl was instrumental in getting the occultation program going at Cornell and always gave it his strong support. In particular, he played an important role in convincing NASA to let us use the KAO for the ε Geminorum occultation.

(*Photo by* Marjorie Nichols)

Peter Usher (left) conversing with Karen and Steve Strom.

COMETS TO QUASARS: SURFACE PHOTOMETRY FROM STANDARD STARS,
AND THE MORPHOLOGY OF THE GALAXY-QUASAR INTERFACE.

Peter D. Usher
Department of Astronomy, The Pennsylvania State University
University Park, PA 16802 USA

Abstract: Wide-angle phenomena in astronomy involve objects
from the Solar System to the farthest reaches of the Cosmos. Pursuit of
knowledge in these areas requires data from cameras of the Schmidt class,
and hence the ability to quantify photographic information. Two fields
of wide-angle research, viz. the photometry of large-scale structures in
comets, and the uranography, taxonomy, and phenomenology, of quasars and
related objects over large solid angles of the sky, are discussed in the
context of Liller's seminal contributions to each. Recent advances in
these areas are: (i) that standard stars can be used to derive and calibrate the surface brightness of extended structures like the Andromeda
galaxy, and hold promise for the determination of a photometric history
of comets past and present; and (ii) that the best data currently
available suggest that $M(B) = -24.0$ is a safe and workable criterion by
which to ensure that quasar samples to $B = 18.5$ mag are comprised wholly
of objects that are unresolved by cameras of the 1.2m Schmidt class.

I. INTRODUCTION.

Liller (1985a) lists astronomical photography as the first of
four major developments that have impelled the advance of modern observational astronomy. Such pre-eminence is deserved owing to the utility of
photography in the solution of an extraordinary diversity of problems.
Sixty years after the discovery (Herschel 1819) that sodium thiosulphate
could 'fix' latent images, astronomers were surprised to discover that
they could photograph the stars. Thereupon, photography became progressively more important in furthering knowledge in the four basic categories
of morphology, photometry, astrometry, and spectroscopy of astronomical
objects.
The virtual permanence afforded by the fixing process has
furthered the study of celestial change in all four of these categories.
For example, the morphology of comet tails were seen to change; photometry begat the period-luminosity relation for Cepheids; astrometry led
to parallax and proper motion; and binary stars gave information on
stellar masses. Observed changes in objects, along with the Cosmological
Principle, ultimately engendered a Worldview that, in broad terms, is the
one held today. The photographic process was paramount in the dramatic
advance of knowledge that culminated in the works of Shapley, Slipher,
Hubble, and their contemporaries.
Despite improved detector sensitivity over the past 20 years,
the reign of photography is presently undiminished in studies of (a)
wide-angle phenomena and (b) archival retrieval. In the former case, no
contemporary charge-coupled device approaches the requisite thousand
million pixels necessary to achieve sub-arcsecond resolution over the
solid angles accessible to Schmidt or Maksutov class cameras. To turn a
phrase, a picture speaks a thousand megawords. Not surprisingly, wide-
field cameras are the workhorses of choice in such programs as the

optical surveys undertaken variously by many workers in their search for statistically complete samples of quasars, galaxies, and stars, to 18th magnitude and fainter (see e.g. Swings 1983, Takase & Miyauchi-Isobe 1987, Philip 1987). In the latter case, the river of time demands perforce that archival records supply the history of past events. The two research areas of this paper exemplify the case. Participants in the Large-Scale Phenomena Network of the International Halley Watch (IHW-LSP) have preserved for posterity an optical record of the morphology of P/Halley during its apparition in 1985 and 1986 (Brandt, Rahe, & Niedner 1982); and 4 remarkable radio-loud quasars studied by W. Liller and co-workers in the Harvard archives [PKS 0537-441 (Liller 1974), PKS 1510-089 and MA 0829+047 (Liller & Liller 1975), and 3C279 (Eachus & Liller 1975)] have astonishingly large variability amplitudes (3.9, 5.4, 3.6, and more than 6.7 mag respectively) that have contributed to the understanding of the role of variability in the distribution of quasars in the Hubble diagram.

Many subfields of interest are subsumed in the categories (a) and (b) above; of these, it is two in particular that I wish to address here, each of which has aspects that belong to each of (a) and (b). §II reviews the problem of the surface photometry of astronomical structures that subtend large solid angles, with a view to the solution of the knotty problem of the photometry of past, present, and (no doubt) future, comets whose images are recorded without sensitometric calibration. Such a goal is in keeping with the theme of this festschrift, for in 1958 W. Liller made the first motion picture of a comet (Miller 1981).

In §III, I shall discuss the problem of the variability of quasars up to that time when their status in the Hubble diagram has been clarified through use of archival and contemporary photographic records, with a view to the solution of the chronic problem (Rees 1986) of what exactly constitutes a 'quasar'. As a result, a safe and workable definition of a '1.2 meter Schmidt camera quasar' will be offered that is slightly more stringent than that of Schmidt & Green (1983). Both of these problems are united by the common theme of W. Liller's research, and the encouragement and support that he has given generously through the years. Moreover, galaxian surface brightness provides a further thematic link between the proximate comets and the distant quasars.

II. SURFACE PHOTOMETRY FROM STANDARD STARS.

A. Statement of the Cometary Problem.

The IHW-LSP network of wide-field cameras did more than chronicle the large-scale morphology of P/Halley in 1985/6; each recorded image is potentially a source of photometric information as well. A worthy goal would be the establishment of a photometrically accurate history of P/Halley through current and past apparitions. The extraction of such data is not easy, for comet images can extend clear across the sky, as was the case for P/Halley in 1910. Type I (plasma) tail structures are ephemeral, often changing in matters of hours. The under-standing of Disconnection Events (DEs; Niedner & Brandt 1978; Niedner 1986), and thus of sector boundaries in the interplanetary magnetic field, is promoted by knowledge both of the morphology and brightness of plasma tails. DEs and other morphologies such as helical wave structures require continual monitoring, but this is often achieved only at large zenith distance, and/or in a twilit sky. Different observers have

different equipment, and sensitometric calibration sources or directional
sky photometers are often lacking. In addition, the major difficulty
with the photographic photometry of extended objects is that photographic
density is not the same as luminous intensity (Mees 1954).
    Brandt (1985) has emphasized that stars of known magnitude
are natural calibrators that should lead to a proper calibration if a
suitable theory is forthcoming, because in principle all the necessary
information is recorded on the emulsion. It is well-known that standard
stars can be used to calibrate other stars, but they cannot be used to
calibrate surface brightnesses in any straightforward way. Moreover, the
goals of morphology and photometry are often antithetical, the former
requiring the best possible definition and hence the narrowest of trailed
star images, the latter benefiting from poor seeing and from star trails
that are as unsaturated as possible.

B. Photometric Methods for Extended Objects.

    Stock & Williams (1962) and de Vaucouleurs (1968) review the
methods of extrafocal images, spot sensitometry, and of schraffier-
kassette and Fabry photometry, and their associated difficulties.
Theories advanced by Kormendy (1973) and Agnelli et al. (1979), and
modifications thereof by Feitzinger et al. (1983), give surface brightnesses calibrated from standard stars, but their reliance on assumptions
concerning star profiles, the need for symmetric profiles and for
simultaneous sky photometry, militates against a general application to
comets.
    The theory of Zou et al. (1981) is independent of assumptions
concerning star profiles, and so is best suited to the (often irregularly) trailed star images in the cometary problem. The H&D curve is
approximated by a polynomial expansion:

$$E_j = \sum_{k=1}^{K} c_k w_j^{kp} \; , \qquad\qquad (1)$$

where W could be the standard dimensionless rectification function of
Baker (1925) and Sampson (1925):

$$W_j = 10^{D_j} - 1 \; ,$$

or the more sophisticated functions proposed by Owaki (1986) to better
account for saturation. Here D_j is the photographic density of pixel j;
c_k are expansion coefficients; and the parameters p,K are described
below.

C. Zero-Point and Extinction for Trailed Standard Stars.

    The theory of Zou et al. (1981) makes no allowance for
differential atmospheric extinction, which is often significant over
large solid angles at large zenith distances. Nor does the theory
incorporate the photometric zero-point, which is inserted ex post facto
by scaling the solution to an assumed sky brightness, as Niedner (1985)
has pointed out. A remedy in theory (Usher 1986) has been implemented

for the Andromeda Galaxy by Speck et al. (1987) (see Fig. 1) with the
help of digitized images reduced to specular photographic densities
(Klinglesmith 1983).

We let E_j in equation (1) be the energy deposited on pixel j
of the emulsion during the exposure time T by a telescope of effective
collecting area A_e . The energy contained in any pixel originates from
both the sky and the object. If the object is a standard star of extra-
atmospheric magnitude B_o and colour $(B-V)_o$, then the energy deposited
from the star must be dimmed by atmospheric extinction, and equated to an
energy that is the difference between the contributions of star plus sky
(* + s) and the sky alone. The latter is estimated from pixels
sufficiently far from the image centre. As a result, for the i = 1, 2,
..., I standard stars:

$$A_e T\, 10^{-0.4[B(0) + B_{o,i} + k'X_i(z) + (B-V)_{o,i}k''X_i(z)]}$$

$$= \sum_{k=1}^{K} C_k \sum_{j=1}^{J} [W_{i,j}^{kp}(*+s) - W_{i,j}^{kp}(s)] , \qquad (2)$$

where B(0) is the zero-point of (say) the Johnson B-band, $X_i(z)$ is the
air mass at zenith distance z, and k' and k" are primary and secondary
extinction coefficients. A solution for k', k", and C_k (k = 1,2,...,K)
may be found by best-fit methods, where I should exceed (K+2) in order
for the system to be overdetermined. The indices p and K are assigned
interactively to produce the best fit to data for a given area of the
plate.

Energy per pixel is converted to intensity of pixel j of an
object of interest "c" according to:

Figure 1: The East-West intensity profile of the Andromeda Galaxy
derived from standard star magnitudes as reported by Speck et al. (1987).
The photometry of Hoag (1952) and de Vaucouleurs (1958) is shown as the
smooth curve.

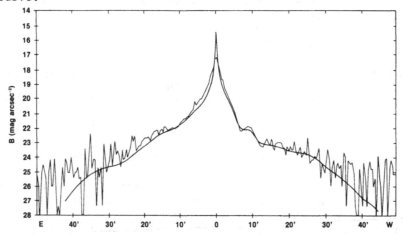

$$I_{o,j}(c) = \frac{\sum\limits_{k=1}^{K} C_k [W_j^{kp}(c+s) - W_j^{kp}(s)]}{\Omega_j A_e T \, 10^{-0.4[k'+k''(B-V)_{o,c}]X_j}} \quad , \qquad (3)$$

where Ω_j is the solid angle of the pixel. The sky brightness of pixel j follows from:

$$A_e T \, 10^{-0.4[B(0)+B_j(s)]} = \sum\limits_{k=1}^{K} C_k \, W_j^{kp}(s) \, .$$

In effect, "s" refers to sky and chemical fog, so the true sky brightness is actually a bit fainter.

In implementing the theory, Speck et al. (1987) use standard stars from the works of Baade & Swope (1963) and McClure & Racine (1969). Plates of M31 were taken by E.P. Moore with the JOCR Schmidt on South Baldy (Brandt et al. 1975), and digitized on the NASA PDS 1010A by D. Klinglesmith. The solution in Fig. 1 agrees with the data of Hoag (1952) and de Vaucouleurs (1958) as nearly as can be ascertained from their figures, except in the very centre of M31 where the galaxy image is saturated. A sky brightness slightly fainter than B(s) = 22.2 ± 0.2 mag per sq. arc second is derived, in good agreement with values elsewhere (de Vaucouleurs 1958; Kalinowski et al. 1975; Pilachowski et al. 1987). The theory shows that the solution for k' and k" is necessary only when there is substantial differential extinction across the field, or when the standard stars have colours substantially different from the object of interest.

The solution must be tested for the case of trailed standard star images and for objects at large z. Liller (1955) has discussed the problem of overlapping stars and nebulae, and there are also problems of overlapping nebulae (e.g. Type I and II comet tails) or galaxies (e.g. M31 and NGC 205). Thus the theory must be even further generalized in order to handle such cases. Liller (1985b) has emphasized the all-important effect of reciprocity failure, which must be modeled in the case of trailed images. The role of the intermittency and fog effects must be elucidated too, for they may act differently for trailed point and extended sources.

III. THE INTERFACE BETWEEN GALAXIES AND QUASARS.

A. The B(max) Hubble Diagram.

If quasars constitute the premier astrophysical puzzle today, how much more enigmatic did their properties seem in the 1970's! While science flourishes on skepticism, those who today defend the possibility that quasars and BL Lacertae objects are not at 'cosmological' distances, are increasingly on the defensive. Fifteen years ago however, the situation was less clear.

A powerful way to investigate the nature of objects of apparent magnitude B and redshift Z is to plot their distribution in the Hubble diagram, B(Z). Setti & Woltjer (1973) set out to explore quasars

in this manner. While they discovered a semblance of a Hubble relation
for radio-selected quasars with canonical radio spectra - implying that
these redshifts at least might be cosmological - they found nothing but
scatter both for radio-selected quasars with flattish radio spectra, and
for optically-discovered quasars. Perhaps for emphasis, or to promote
levity in an otherwise serious subject, one data point (3C273; their Fig.
1b) fell <u>outside</u> the borders that they had drawn around the scatter! The
authors were consoled only by the fact that all quasars were situated
above the standard candle line for elliptical galaxies, a fact that
Sandage (1972) had previously asserted in connection with evidence that N
galaxies housed mini-quasars in their nuclei.

 Following the identification (Schmitt 1968) of the variable
star BL Lac (Hoffmeister 1929) with the radio source VRO 42.22.01 (McLeod
et al. 1965), Strittmatter et al. (1972) proposed that objects exhibiting
the properties of BL Lac belonged to a discrete class of quasar. Among
unusual properties that were necessary - if not sufficient - for their
classification (Stein et al. 1976; Wolfe 1978), were rapid optical
variability, the absence of emission lines, and flat or inverted radio
spectra. Clearly, variability would render apparent brightness uncertain
in the $B(Z)$ Hubble diagram, even if redshifts were known. Although a few
objects that warranted the BL Lac classification had known Z, and still
others were helpfully imbedded in nebulosity, the problem of optical
variability of the starlike component remained. Photometric histories of
all types of quasar were therefore a minimum condition for progress.

 In 1974, W. Liller afforded J.T. Pollock the opportunity to
use Harvard's archival collection in order to extract the necessary
photometric histories of BL Lac objects. Advances reported here and
elsewhere stem from the generosity and scientific insight of W. Liller 13
years ago. Among other results was the quantification of the correlation
between two seemingly disparate properties of the BL Lac class, viz. the
optical amplitude of variability $\Delta B = B(max)-B(min)$, and the 5 GHz radio
spectral index (Pollock 1975a, 1975b). The results for a significant
sample of BL Lac objects have proved useful not just in establishing the
consistency of optical-radio relationships for all members of the class,
but thereby by inference establishing the lower bound to the quasar
distribution in the Hubble diagram for all members on the strength of the
redshifts for a few. The latter was accomplished with the help of the
apparent magnitude at minimum brightness $B(max)$ (Pollock 1975a, Usher
1975), later bolstered by the signal-to-noise ratio Q of ΔB (Usher 1978).

 $B(max)$ made up in astrophysical significance what it lacked
in statistical robustness; for though it was an extremum statistic, it
was often substantiated by the entire archival history, and became
therefore a fairly secure measure of the faintest luminosity attained by
the quasar component. Indeed, when in the course of its variation, the
quasar component makes no contribution, $B(max)$ should in theory be set by
the luminosity of the putative host galaxy itself. The Hubble relation
would then be sought in the Hubble diagram of $B(max)$ vs. Z, rather than
$B(Z)$. A lower bound to the distribution of quasars with a broad
luminosity function should then be close to that for the Sandage galaxy
line if redshifts were cosmological.

 It is noteworthy that extremum statistics have long played a
role in quantifying observed limits to physical processes, such as those
of Eddington or Chandrasekhar. Statistics of all kinds are essential to
epistemology, for they are the grounds of knowledge in the experimental
sciences, and particularly in the observational science of astronomy.

They promote confidence in the interpretation of the book of nature by measuring progress toward truth and understanding. Statistics are the vehicles of communication between the global and the particular, as each plays its clarifying role in the seemingly endless oscillations of the Hermeneutic Circle (e.g. Usher 1980).

When the subset of BL Lac objects having known Z was seen thus to conform, and since all BL Lac objects showed self-consistent optical-radio properties, the case for their extragalactic nature was strengthened. The case could be said to be proved but for the chance of an astrophysical conspiracy. Still, subsequent studies to fainter survey

Figure 2: Morphological type determined from Palomar 1.2 m Schmidt camera plates or prints, is shown on the Hubble diagram for objects selected in the PG and US surveys. Filled and open circles denote unresolved and resolved objects respectively, and half-filled circles denote possibly resolved objects. Dots represent objects with otherwise uncertain photometry and/or unknown morphology. The slanted curves are luminosity criteria (H_0 = 50 km/sec/Mpc, q_0 = 0); the lower is the c-model of Bruzual (1983) with no renormalization, and the upper is M(B) = −24.0 for a power law frequency spectrum of index −0.5 (Schmidt & Green 1983). The horizontal line is the cut made by the U-B guillotine.

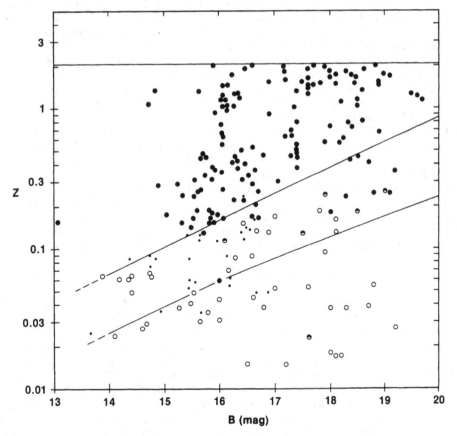

limits of all known radio- and optically-selected quasars with
variability statistics (Usher & Mitchell 1978) showed that the B(max)
Hubble diagram was less scattered, thereby in turn diminishing the
difficulties encountered by Setti & Woltjer (1973). It seemed reasonable
that quasars follow the Hubble expansion if they have a broad luminosity
function, regardless of their radio properties.

B.   The Quest for Complete Samples.

These advances were nonetheless beset by the unavailability
of statistically complete samples, not to mention an archival photometric
history of each member. Optically selected quasars outnumber their radio
counterparts by large factors (Sandage 1965), and thus photographic
surveys for bright samples predominate. To date, selection by ultraviolet colour excess (UVX) has been most popular owing to its apparent
success in identifying complete samples of quasars up to the Lyα
"guillotine" at Z = 2.1 (when Lyα is redshifted out of the UV; Fig 2),
while prism/grism surveys have attempted to achieve completeness at
higher Z. Selection effects are paramount in all cases, for magnitude
limited samples are only as complete as their selection effects are known
and quantified.

For this reason, selection effects have been systematically
quantified in the US survey for blue- and ultraviolet-excess quasars
(Huang & Usher 1984 and references therein). These surveys have been
made in 6 fields selected by Sandage (1976) that contain photometric
standards measured by him and by Purgathofer (1969). Of the many
selection effects that obtain, 4 may be seen in relation to the 2 limits
in each of the 2 dimensions of the Hubble diagram (cf. Fig. 2). The
bright limit in apparent magnitude B is set by the distribution of
quasars closest to the Milky Way; selection effects at this limit have
been investigated by Warnock et al. (1986). The faint limit in apparent
magnitude B is the limit accessible to observation. The upper-left and
lower-right limits to the distribution may be regarded as luminosity
limits, and as such are functions of both B and Z; they are best
described in terms coined by Sandage (1972): (i) in colour surveys that
use the (U-B) colour index, the upper limit is not solely the natural one
set by the extreme luminosity of quasars, but is often subject to a
relatively clean cut made by the U-B "guillotine", as seen in Fig. 2;
(ii) whereas it is the lower luminosity limit that is pertinent here, for
it leads to the "twilight zone" between galaxies and quasars.

C.   Exorcising the Twilight Zone.

The investigation of the twilight zone is essential to the
solution of the tiresome problem of morphological selection of quasar
candidates in photographic surveys, and thus ultimately to the determination of the sort of universe we live in. Morphology comes into play
whenever a quasar candidate is selected, for how "quasi-stellar" must a
"quasi-stellar object" be to qualify for selection? The US survey, for
instance, specifically selects against resolved objects.
        Fig. 2 contains most known data from the Palomar-Green (PG)
survey (Schmidt & Green 1983), from the US survey (Mitchell et al. 1984),
and from a related UVX galaxy survey (Mitchell et al. 1982). It also
embodies data from Bahcall et al. (1969), Burbidge & Strittmatter (1972),
Peterson & Bolton (1972), Wills & Wills (1976), Browne & Savage (1977),

Kron (1985), Wampler & Ponz (1985), Mitchell & Warnock (1986), and Green & Usher (unpublished).

The PG and US surveys have a common selection criterion for morphological type, which is judged by eye from Palomar 1.2m Schmidt camera images. Under such conditions, Kristian (1973) has shown that the quasar component must be progressively less luminous if fuzz is to be seen around objects at increasingly large redshifts. The perceived border between resolved and unresolved objects therefore would not have a slope comparable to that of a standard candle line like the two curves shown in Fig. 2 (and described in its caption). Rather, the slope would be much flatter. Inspection of Fig. 2 shows that resolved and unresolved objects (filled and unfilled circles) divide roughly along a curve with a flattish slope that cuts from left to right between the 2 lines in Fig. 2, as Kristian's theory would predict. The position of this line of demarcation can be fit to the data of Fig. 2 for reasonable values of the parameters in Kristian's theory, and with the help of the magnitude calibration of POSS stellar image sizes given by Liller and Liller (1975). The equation is:

$$B(mag) = 22.57 - 0.850 \ (C/Z) + 0.0126 \ (C/Z)^2 \ . \tag{4}$$

The fit parameter C depends inter alia on the capabilities of the surveyor's eye. From Figure 2, C is roughly 1.5, suggesting a somewhat better than average ability to distinguish resolved and unresolved objects. This may be accounted for in the US survey at least in part by the fact that the edge of the twilight zone was judged not just by the visibility of fuzz surrounding starlike objects, but also by an appearance of 'softness' to apparent point sources on original negatives.

The semi-empirical theory of Kristian, which is consistent with galaxies (and presumably their bright semi-stellar nuclei too) being at their cosmological distance, is in accord with the present data. This suggests that a Kuhnian crisis is not yet at hand; pursuant to normal science, bona fide anomalies may be sought through further puzzle-solving (Kuhn 1970).

For uranographic purposes then, a safe and workable criterion is necessary in order to characterize the resolvability of compact objects seen through the eyes of a 1.2m Schmidt camera and its users. A criterion such as equation (4) is not particularly convenient to use however; a pure luminosity criterion is more desirable. The upper curve in Fig. 2 suggests that a luminosity of M(B) = -24.0, defined in the manner of Schmidt and Green (1983), safely separates stellar from nearly-stellar and non-stellar objects (except possibly at the very brightest apparent magnitudes). This is a semi-empirical "1.2 m Schmidt criterion" for morphological selection, good to the US survey limit of about B = 19.0 mag. The twilight zone inhabited by BL Lac objects, and Seyfert, N, and E galaxies, is thereby cleanly excised to the stated limits, and the practical definition of a 'quasar' is rendered more precise (Schmidt 1969; Rees 1986).

This work was supported in part by NASA Grant NAG 5-361.

References:

Agnelli, G., Nanni, D., Pitella, G., Travese, D. & Vignato, A. (1979).
        Ast. & Ap. 77, 45.
Baade, W. & Swope, H. (1963). A. J. 68, 435.
Bahcall, J.N., Schmidt, M. & Gunn, J.E. (1969). Ap. J. Letters 157, L77.
Baker, E.A. (1925). Proc. R. Soc. Edin. 45, Part II, 166.
Brandt, J.C. (1985). Private communication.
Brandt, J.C., Colgate, S.A., Hobbs, R.W., Hume, W., Maran, S.P.,
        Moore, E.P. & Roosen, R.G. (1975). Mercury 4, No. 2, 12.
Brandt, J.C., Rahe, J. & Niedner, M.B. (1982). International Halley Watch
        Newsletter No.1, ed. S.J. Edberg, p. 13. NASA-JPL 410-5 8/82.
Browne, I.W.A. & Savage, A. (1977). MNRAS 179, 65P.
Bruzual, A.G. (1983). Rev. Mex. Ast. Astrof. 8, 63.
Burbidge, E.M., & Strittmatter, P.A. (1972). Ap. J. Letters 174, L57.
de Vaucouleurs, G. (1958). Ap. J. 128, 465.
Eachus, L.J., & Liller, W. (1975). Ap. J. Letters 200, L61.
Feitzinger, J.V., Nicolov, A., Schmidt-Kaler, T., & Tennigkeit, J.(1983).
        Ast. & Ap. 126, 352.
Herschel. J.F.W. (1819). Edin. Phil. J. I, 8 & 396.
Hoag, A.A. (1952). Thesis, Harvard College Observatory.
Hoffmeister, C. (1929). Astron. Nachr. 236, 233.
Huang, K-L., & Usher, P.D. (1984). Ap. J. Suppl. 56, 393.
Kalinowski, J.K., Roosen, R.G., & Brandt, J.C. (1975). Publ. Ast. Soc.
        Pac. 87, 869.
Klinglesmith, D.A. (1983). Ed., Proc. Astron. Microdensitometry Conf.,
        NASA-Goddard Space Flight Center. NASA Conf. Publ. #2317.
Kormendy, J. (1973). A. J. 78, 255.
Kristian, J. (1973). Ap. J. Letters 179, L61.
Kron, R. (1985). Private communication.
Kuhn, T.S. (1970). The Structure of Scientific Revolutions, 2nd ed.,
        Chap. VII. Chicago: Univ. of Chicago Press.
Liller, M.H., & Liller, W. (1975). Ap. J. Letters 199, L133.
Liller, W. (1955). Ap. J. 122, 240.
Liller, W. (1974). Ap. J. Letters 189, L101.
Liller, W. (1985a). In The Cambridge Astronomy Guide: A Practical
        Introduction to Astronomy, p. 15. Cambridge: Cambridge
        University Press.
Liller, W. (1985b). Report on the Meeting of the Steering Committee of
        the International Halley Watch Large-Scale Phenomena,
        NASA-Goddard Space Flight Center, Greenbelt, MD, May 29.
McClure, R.D., & Racine, R. (1969). A. J. 74, 1000.
McLeod, J.M., Swenson, G.W., Yang, K.S., & Dickel, J.R. (1965).
        A. J. 70, 756.
Mees, C.E.K. (1954). The Theory of the Photographic Process, p. 162.
        New York: Macmillan.
Miller, F.D. (1981). Proc. Workshop Modern Observational Techniques for
        Comets, NASA-Goddard Space Flight Center, October 22 - 24,
        p. 169. NASA-JPL 81-68.
Mitchell, K.J., Brotzman, L.E., Warnock, A., & Usher, P.D. (1982).
        Ast. Sp. Sci. 88, 219.
Mitchell, K.J., & Warnock, A. (1986). Private communication.
Mitchell, K.J., Warnock, A., & Usher, P.D. (1984). Ap.J. Letters 287, L3.
Niedner, M.B. (1985). Private communication.
Niedner, M.B. (1986). International Halley Watch Newsletter No. 9, ed.
        S.J. Edberg, p. 2. NASA-JPL 410-20 10/86.

Niedner, M.B., & Brandt, J.C. (1978). Ap. J. 223, 655.
Owaki, N. (1986). Observatory 106, 194.
Peterson, B.A., & Bolton, J.G. (1972). Ap. J. Letters 173, L19.
Philip, A.G.D. (1987). Ed., I.A.U. Colloquium No. 95, The Second
        Conference on Faint Blue Stars, Tucson AZ, 1 - 5 June.
Pilachowski, C., Africano, J., Goodrich, B., & Binkert, W. (1987).
        Nat. Opt. Astro. Obs. Newsletter No. 9, p. 8.
Pollock, J.T. (1975). Ap. J. Letters 198, L53.
Pollock, J.T. (1975). M.S. Thesis; Penn. State Univ. Radio Astron. Obs.
        Sci. Report 029.
Purgathofer, A.T. (1969). Lowell Obs. Bull. No. 147, Vol. VII, 98.
Rees, M.J. (1986). International Astronomical Union Symp. No. 119, 1.
Sampson, R.A. (1925). MNRAS 85, 212.
Sandage, A. (1965). Ap. J. 141, 1560.
Sandage, A. (1972). Ap. J. 178, 25.
Sandage, A. (1976). Private communication.
Schmidt, M. (1969). Ann. Rev. Astron. Astrophys. Vol. 7, 527.
Schmidt, M., & Green, R.F. (1983). Ap. J. 269, 352.
Schmitt, J. (1968). Nature 218, 663.
Setti, G., & Woltjer, L. (1973). Ap. J. Letters 181, L61.
Speck, S.F., Usher, P.D., Klinglesmith, D.A., & Niedner, M.B. (1987).
        Paper presented at the 170th Meeting of the American
        Astronomical Society, Vancouver BC, June 14 - 17.
Stein, W.A., O'Dell, S.L., & Strittmatter, P.A. (1976).
        Ann. Rev. Ast. & Ap. 14, 173.
Stock, J., & Williams, A.D. (1962). In Astronomical Techniques, ed.
        W.A. Hiltner, p. 374. Chicago: Univ. of Chicago Press.
Strittmatter, P.A., Serkowski, K., Carswell, R., Stein, W.A., Merrill,
        K.M., & Burbidge, E.M. (1972). Ap. J. Letters 175, L7.
Swings, J-P. (1983). Ed., Proc. 24th Liege Astrophysical Colloquium,
        Quasars and Gravitational Lenses, Liege, Belgium, June 21-24.
Takase, B. & Miyauchi-Isobe, N. (1987). Ann. Tokyo Ast. Obs.,
        2nd Series 21, 251.
Usher, P.D. (1975). Ap. J. Letters 198, L57.
Usher, P.D. (1978). Ap. J. 222, 40.
Usher, P.D. (1980). Astron. Quart. 3, 178.
Usher, P.D. (1986). Proc. SPIE 702, 387.
Usher, P.D., & Mitchell, K.J. (1978). Ap. J. 223, 1.
Wampler, E.J., & Ponz, D. (1985). Ap. J. 298, 448.
Warnock, A., Usher, P.D., Mitchell, K.J., & Howell, S.B. (1986).
        MNRAS 218, 445.
Wills, D., & Wills, B.J. (1976). Ap. J. Suppl. 31, 143.
Wolfe, A.M. (1978). Ed., Pittsburgh Conference on BL Lac Objects,
        April 24-26. Pittsburgh: University of Pittsburgh.
Zou, Z-L., Chen, J-S., & Peterson, B.A. (1981). Chin. Ast. & Ap. 5, 316.

Questions:

Leif Robinson: Why don't the dark lanes in the intensity map of M31
                match up with dark lanes in direct photographs?

Usher: The pictures shown do not have the resolution to which we are
         accustomed. For this particular research, morphology is
         secondary to photometry. Apart from seeing and focus, image
         resolution in this case is affected by four factors arising
         from instrumental and display capabilities: the JOCR Schmidt
         has a plate scale of nearly 300"/mm, over 4 times greater
         than the Palomar 1.2 m Schmidt; the original image on a IIa-O
         emulsion has been digitized with a resolution of 6"/pixel;
         the resolution of Speck's false-colour slide has been further
         degraded by a factor of 4 in order to fit the image on the
         CRT screen; and the resolution of the slide shown is again
         diminished by use of only 10 discrete hues to encompass the
         entire range of surface brightness from the nucleus to the
         sky. A dark lane must be wider than about 24" to be seen,
         and in fact the first lane due west of the nucleus, which is
         roughly 35" wide, is just detectable in Fig. 1.

Yervant Terzian: Does your magnitude-redshift diagram of quasars follow
                  the Hubble Law?

Usher: The Hubble diagram in Fig. 2 is not yet in a form suitable for
         modeling, since it is being used here to investigate morpho-
         logical classification of selected objects. The major
         corrections needed in Fig. 2 are: the addition of complete
         samples of high redshift quasars from surveys that are not
         subject to the U-B guillotine; the exclusion of resolved
         objects with the help of a self-consistent theory; and the
         use of magnitude-limited samples. These would alter the
         appearance of the diagram quite considerably. Whereupon,
         problems of photometric error, variability, confusion, over-
         sight, and so on, must be addressed for each sample used.

(*Photo by* Marjorie Nichols)

Jay Pasachoff delivering his talk.

OBSERVING SOLAR ECLIPSES

Jay M. Pasachoff

Williams College–Hopkins Observatory, Williamstown, Massachusetts 01267

Total solar eclipses are the most dramatic phenomena that can be seen on earth. The blotting out of the sun in the middle of the day has drama for all humanity. However, this celestial phenomenon is rare–on average, a total eclipse occurs every year and a half somewhere on earth (Fig. 1). If you stay at one location on earth, it is usually more than 300 years before another eclipse occurs. Total solar eclipses provide valuable information about the upper layers of the sun–the chromosphere and corona, in particular. Though data from eclipses can now be supplemented by data from certain spots on earth or, when a suitable spacecraft is aloft, by satellite data, these other observations are supplemental to the eclipse data. The data from all these sources taken together are improving our understanding of the solar atmosphere.

ECLIPSES OF THE CURRENT SAROS
1973

The duration of an eclipse depends on the relative positions of the sun, moon, and earth in their syzygy. Primarily because of the elliptical orbit of the moon around the earth, the moon's disk is sometimes smaller than the sun's disk–an annular eclipse. Sometimes the moon's disk barely covers the sun's disk, a circumstance that leads to an eclipse with only fleeting instants or seconds of totality, as in the 1987 eclipse that was observed only from an airplane off the coast of Iceland. At other times, an eclipse can last as long as 7 minutes. Such a long eclipse crossed Africa on June 30, 1973.

At this Liller Symposium, I am glad to report that Bill Liller saw fit to organize active teams to study this long 1973 eclipse. His teams and my own were part of the U.S. national expedition, sponsored by the National Science Foundation. The locations where totality exceeded 7 minutes were to be in the Sahara Desert in Mali and Niger, and Donald Menzel and I had visited Timbuktu and Agadez in 1970 to investigate prospective sites. Supplying an eclipse expedition in the desert 100 km north of these fabled cities appeared then to be exceedingly difficult, and the large amounts of dust that were placed in the air by the drought that followed over the next years made these 7-minute sites less desirable.

After much consideration, the NCAR staff who ran the eclipse expedition for the NSF decided that the main body of observers should go to the Northern Frontier District of Kenya, on the shore of Lake Rudolf (now Lake Tanganyika). Though totality would last only 5 minutes, and there was a somewhat higher chance of clouds, we would be relatively free of dust. Only people who justified a particularly long totality went to Mauritania, where totality was to exceed 6 minutes. As it turned out, Bill Liller took one of his teams to Mauritania, and sent the Matteis and others on the team to Kenya. I went with my assistants and students to Kenya

(Fig. 2), though I also sent a student to work with Donald Menzel in Mauritania. The National Geographic Society supported our joint work.

During the event, a dust storm came up in Mauritania, cutting the transparency there to about 10 per cent. In Kenya, we survived clouds that crossed the sun only 10 minutes before totality, and observed totality in very clear sky (Fig. 3). My experiment was to study the forbidden lines of twelve-times-ionized iron; they are the strongest coronal lines in line-to-continuum ratio, yet are little studied because they are located in the infrared at 10,747 Å and 10,798 Å. Since one is radiatively and the other collisionally controlled, their ratio is a sensitive measure of electron density.

1973 Annular
On December 24, 1973, Comet Kohoutek was near perihelion during the annular eclipse. I tried to observe the comet during the eclipse from a site high in the Andes above Bogota, Colombia, but the sky remained too bright to see the comet (Fig. 4). Only the Skylab astronauts were able to see the comet during its perihelion.

1974
An eclipse crossed Western Australia on June 20, 1974, the Australian winter. We observed from a site near Albany, Western Australia, a part of Australia recently made famous by the America's Cup. Though we had a streak of unprecedented good weather for that time of year in the week preceding the eclipse, eclipse day was mostly cloudy. To see the eclipse, my wife and I had to go up in a small plane at the last minute. Unfortunately, the main eclipse site was on the landing field, so we had to have the plane land on a field nearby; as a result we barely got high enough to see totality.

Unfortunately, an excellent experiment of the Australian team was wiped out by clouds. They were unable to get the independent measure of the electron-temperature of the corona they were seeking by studying photoelectrically the extremely broadened and overlapping Fraunhofer lines in the coronal spectrum. Further, an attempt to use a rocket to study the Lyman-alpha line in the corona was unsuccessful, as both rockets were lost.

1976
An eclipse crossed eastern Australia on October 23, 1976; I was unable to attend. Some sites were cloudy, including the one where I would have been.

1977
The October 12, 1977, eclipse crossed no land, and was visible only from mid-Pacific. I arranged to work with Los Alamos scientists, who brought two inertial guidance platforms aboard the T.S.S. Fairsea. Unfortunately, clouds blanketed the Pacific, though satellite data radioed to the ship enabled the Fairsea and its sister ship to travel rapidly

through the night to the place where the clouds would begin to break. We succeeded in observing the eclipse through a hole in the clouds, and obtained the first good image of the corona in one of the iron-thirteen lines (Fig. 5). But the sky was not of photometric quality, so we could not make our comparison of the lines in the doublet.

1979
The February 26, 1979, eclipse crossed the Pacific Northwest of the United States and much of middle and northern Canada. My group observed from Brandon, Manitoba. We repeated our spectroscopic experiment, using versions of vidicons from both Tektronix and Princeton Applied Research.

The day of the eclipse was good enough for spectroscopy, though not completely clear (Fig. 6).

1980
A major eclipse of favorable circumstances crossed Africa and India on February 16, 1980. The U.S. team, sponsored by the NSF with NCAR logistics, went to the Japal-Rangapur Observatory south of Hyderabad, India. My group continued our spectroscopic studies of the forbidden iron-thirteen lines but also started a new investigation related to coronal heating. Studies from the OSO-8 spacecraft had recently discovered that the acoustic flux passing through the chromosphere was insufficient by a factor of 1000 to heat the corona to its observed temperature of 2,000,000 K. The old theory of acoustic-wave coronal heating had to be discarded. New theories involved coronal magnetic or electric fields; in any case, the x-ray pictures of the sun taken from Skylab and from rockets had revealed that the corona was hottest over active regions, where the magnetic field was most important. My own experiment, which was joint with Donald A. Landman (then of the Institute for Astronomy of the University of Hawaii and now of Mission Research, Inc.), was to search for high-frequency oscillations of the corona, with a period of perhaps 1 Hz. The equipment, designed by J. Phil Schierer (now of Photon Kinetics, Inc., and then of Tektronix, Inc.), used fiber optics to bring small regions of the corona to cooled photomultipliers. Bruce Miller (now of Bipolar Integrated Technologies and then of Tektronix), designed and built the digitizing electronics.

The weather for the eclipse turned out to be fine (Fig. 7), and our equipment functioned well. Landman and I reported (*Solar Physics* 90, 325-330, 1984) the presence of excess power in the 0.5-2 Hz range.

1981
The July 31, 1981, eclipse was visible from the ground near Lake Baikal in the U.S.S.R. I observed the end of the eclipse from an airplane off the coast of Kauai, Hawaii (Fig. 8). Split-second timing was necessary; a small plane following our larger plane apparently missed its turn by a few seconds and missed totality!

1983

The major eclipse in Java, Indonesia, on June 15, 1983, was the subject of another NSF-sponsored expedition. My group repeated our coronal-heating experiment, but with second-generation equipment that was much improved in a variety of ways. We had new, sturdier optical fibers; better methods of changing the location of fibers during totality; better tracking (Pasachoff and W. Livingston, *Applied Optics* **23**, 2803-2808; and M. Demianski and Pasachoff, *Solar Physics* **93**, 211-217), and better digitizing.

Though a dozen U.S. teams, and neighboring Indian teams, were set up on the north shore of East Java at Tanjung Kodok, it poured rain the night before the eclipse. The sky finally cleared sufficiently to observe the eclipse, but it was far from photometric quality (Fig. 9).

With additional support from the National Geographic Society, I also teamed with Jim Elliot of MIT to use his photoelectric equipment to observe first the eclipse and then, a few days later, the occultation of a star by Neptune in a search for rings. Though we set up at the Bosscha Observatory in Bandung, we were clouded out of the occultation.

We were closer to solar minimum than at the 1980 eclipse, so the corona had the asymmetric shape typical of that phase of the solar activity cycle.

Our coronal-heating experiment (Pasachoff and E. F. Ladd, *Solar Physics* **109**, 365-372, 1987) gave excellent results, and we again found excess oscillatory power in the .5-2.0 Hz range in a coronal loop. This time we had additional monitoring to make certain that the oscillations were solar rather than in the terrestrial atmosphere.

1984 annular

An annular eclipse that was 99.9 per cent total crossed Mexico and the United States on May 30, 1984. Clouds covered eastern Virginia, but sites from Atlanta westward were clear. The skies were extremely clear at my site in Picayune, Mississippi, just north of New Orleans. The corona could even be seen briefly. The eclipse was so close to total that an unbroken annulus of sunlight never appeared; it was always broken by mountains on the moon.

1984

The November 1984 eclipse was visible in Papua New Guinea in the very early morning. Since totality was to be only 40 seconds, and the sun was only to be 23 degrees in altitude, few observers went. I observed from a site near the Hula peninsula, and the weather turned out to be perfect (Fig. 10). It was the clearest sky I had seen at an eclipse since the Mexican eclipse of 1970, and it was truly a pity that all the large equipment that had been in Java in 1983 was not present. After all, even though totality was 40 seconds instead of 5 minutes, all the phenomena of the diamond ring, chromosphere, and prominences can be seen as well at a shorter eclipse, and 40 seconds is ample time to study 1 Hz oscillations or to take a number of exposures.

A timing study we made, using a video camera, has given a public assessment (Pasachoff and B. O. Nelson, *Solar Physics* **108**, 191-194, 1987) of the accuracy of timing predictions. The extent of the discrepancy between the times for second and third contact that we observed and the predictions, even corrected and updated soon before or after the eclipse, was such that I do not believe the validity of the previously reported conclusion that a secular change in the diameter of the sun had been established by eclipse observations over several decades. Since a monotonic secular change of the reported magnitude would have significant effects on terrestrial climate, this type of study should be continued.

We are still reducing our calibrated data on the shape of the corona. John MacKenty (formerly of the Institute for Astronomy of the University of Hawaii and now of the Space Telescope Science Institute), has worked with me to make a radial-filter algorithm. By appling the algorithm to the eclipse data, we succeeded in studying the eclipse to 4.5 solar radii. We are tying in our eclipse data with features observed on the solar surface or as recorded on the Mauna Loa coronagraph and provided by D. Sime (HAO/NCAR).

1988

The March 17-18, 1988, eclipse crossed Sumatra and Borneo in Indonesia, and then the southern Philippines. I went with a student to Sembawa, toward the centerline from Palembang, South Sumatra. Though we saw first contact, a fog arose about 15 minutes before totality and obscured our view until after third contact. We later saw some partial phases and fourth contact. Many observers at other sites in Indonesia and the Philippines did see totality, though some others were clouded out.

The Kuiper Airborne Observatory flew out of Guam, and scientists on board successfully observed at wavelengths of 200 μm, 400 μm, and 800 μm. An x-ray rocket experiment made images of the sun at 175 Å on the day of the eclipse. The HAO radial-filter telescope operated in its "clouds present" mode, making a radial-filter image but not taking polarization data.

THE CASE FOR ECLIPSES

Several questions about eclipses are often asked by students and by the general public, and I shall deal with these common queries here.

Why Study the Solar Corona?

Though the solar photosphere is best visible from the ground every clear day, the solar corona is visible from the ground at eclipses. Scientists are interested in the corona for four major categories of reasons. First, we learn about the sun itself, how it works. The question of how the sun shines is of great general interest, and the specifics of the origin of the radiation that reaches us from the corona is one of the detailed phases of the general question. All the energy that leaves the sun passes into or through the chromosphere and corona. Why the corona temperature is as high as 2,000,000 K is still a major question, with theories of coronal heating via magnetic fields currently dominating.

Secondly, we study the solar corona to learn more about the earth. The corona is continually expanding into space in the form of the solar wind, which envelops the earth. At an eclipse, we see the corona only for the first few million kilometers surrounding the solar photosphere. But the outer parts of the corona extend over 150,000,000 km, beyond the orbit of the earth. The sun affects many things on earth. For example, it leads to aurorae, short-wave and CB communication blackouts (with both civilian and military consequences), and surges on power lines that can lead to outages. Many scientists are increasingly coming to believe that changes of the solar radiation affect the earth's climate on a time scale that may be detectable by us. Indeed, we are now waiting for the next solar activity maximum to be over and the decline to start in order to assess the overall drift of the solar constant that has been recorded by the ACRIM instrument aboard the Solar Maximum Mission since the beginning of this decade. So an understanding of the sun is necessary to understand the earth itself.

Third, the sun is but a typical star, one of brightness and luminosity that are in the middle of the possible range of these parameters. The phenomena that we are studying on the sun exist on many other stars as well, so by studying the sun we are gaining detailed closeup information that we can extend to more distant stars. The study of the sun is thus an important part of understanding the universe. Studies with the International Ultraviolet Explorer satellite, for example, have allowed us to study spectra from stars that show transition zones between stellar chromospheres and coronae. The Einstein Observatory and more recent x-ray spacecraft have allowed us to observe x-rays from the stellar coronae themselves. Stars other than the sun have more powerful coronae than we had expected on the basis of solar observations. Further studies of the sun will show us ways to resolve the discrepancy.

Fourth, the sun is a celestial laboratory that we use to learn basic laws of physics. The conditions that we find on the sun are often not duplicable in laboratories on earth. For example, the density of the solar corona is so low that it would be considered a fantastic vacuum in a laboratory on earth. The corona is a hot plasma which the solar magnetic field directs into the beautiful streamers that we see at eclipses. Thus it seems clear that detailed study of the corona will help us discover basic laws of physics that explain the action and motion of hot gas held in a magnetic field.

This type of knowledge is useful in the study of magnetic-fusion research to provide energy on earth. On earth, we have trouble holding plasma together sufficiently long for protons and deuterons to come close enough together to fuse, releasing energy following Einstein's equivalence $E=mc^2$. The tendency of these positively charged particles to repel each other is strong. In a hot gas, particle velocities are sufficiently high for protons and deuterons to approach each other closely in spite of their mutual repulsion. But no material container can hold such hot gas, so magnetic fields are used to constrain the gas. At Princeton and elsewhere, progress has been made in constraining the gas at high temperature and increasing the length of time the gas can be constrained. The kinds of laws of physics we learn from studies of the corona are just the kind that will help us with this important problem.

This is not to say that the astronomers are the ones who will actually apply their discoveries to practical ends. The astronomers are, for the most part, carrying out their observations because they are excited by the astronomical projects or by the physical laws. It will be others, perhaps working decades in the future, who will apply the astronomical discoveries. There is a history, dating back hundreds of years, of astronomical discoveries being at the base of later technological development. I am sure that it will happen here, too.

Why Study Eclipses from the Ground

Several solar telescopes have been placed in orbit, including some on the Skylab Missions in 1973-74, a Defense Department satellite that was in orbit from 1978 until it was shot down in an SDI trial in 1985, and the Solar Maximum Mission, launched in 1980. Solar telescopes flew aboard Spacelab 2 on a space shuttle. The set of observations provided has been magnificent and provided new insights into many fields of solar physics. But they by no means answered all the existing questions. Much remains to be learned, and no possible sources of answers--including eclipses--should be ignored.

Further, though some of the telescopes in space can observe the sun in parts of the spectrum that do not reach the ground, in other ways their observations are inferior to eclipse observations. In space-borne coronagraphs, not only the solar photosphere but also a zone of the corona around the photosphere has to be occulted, to limit scattering of sunlight in the telescope. Thus the inner corona is hidden. The size of the occulting disk on the Solar Maximum Mission is 1.75 solar radii (Fig. 11).

Ground-based observations of the corona outside the eclipse are limited in other ways. Most coronagraph images can see only the innermost corona (Fig. 12), often only in one of a few coronal spectral lines, and cannot see the corona as far from the sun as can be seen at an eclipse. The Mauna Loa coronagraph of the High Altitude Observatory uses polarization to see the corona somewhat farther out (Fig. 13).

At an eclipse, one can carry out a broad range of studies, including spectroscopy, through the lower and middle corona. The eclipse observations are necessary along with the ground-based coronal observations and the space-based coronal observations to get the full picture. For example, our observations from Papua New Guinea of the total solar eclipse of 1984 gave observations of the corona from the solar limb out to 4.5 solar radii, as shown on an image that John MacKenty and I have analyzed with a radial-filter algorithm (Fig. 14).

Also, continual advances in theory and in equipment allow new observations to be formulated and new equipment to be provided on short notice for an eclipse expedition. Eclipse equipment, unlike equipment to be launched into space, does not have to be of launch standards of sturdiness.

The traditional advantage of eclipses over space observations remains–namely, the comparative ease and low cost of transport of state-of-the-art equipment and very bulky items to eclipse sites. And eclipse equipment can be mounted on sturdy bases, with the earth as a platform, and be adjusted and tuned at the last minute by qualified scientists. The advantages of this are considerable. It is not often realized, for example, that the jitter in spacecraft and limitations on the size of spacecraft telescopes have meant that no spacecraft have yet been able to study details on the sun the size of the finest details that have been studied from earth.

Many scientific problems can be studied from the ground as well as or better than from space. A whole realm of observations of the solar chromosphere and corona in the optical and near-infrared parts of the spectrum can and should be done from the ground at eclipses rather than from space. Just as new 8-m and 10-m telescopes are being built on the ground, and will be given even more work by the launch of the Hubble Space Telescope, solar telescopes on the ground not only have lots to do but will have even more to do when new solar telescopes are aloft.

Further, spacecraft are few and far between. The Solar Maximum Mission, launched in 1980, ended in 1989. The Solar Optical Telescope was downsized to be the High Resolution Solar Observatory, which, in turn, was put on hold and has been resurrected as the Orbiting Solar Laboratory. We wait for it and for a Spacelab with solar instruments, neither of which will fly for many years. So the hope of the solar-physics community of having high resolution from space will not happen soon.

The contrast in cost between space observations and ground-based observations is extreme. The Skylab solar telescopes and the Solar Maximum mission cost hundreds of millions of dollars each. Even the repair of the Solar Maximum Mission (Fig. 15) cost over a hundred million dollars. For less than one-tenth of one per cent of these costs, one can send a well-equipped modern ground-based eclipse experiment. Even allowing for some of the eclipses to be clouded out, eclipses are a very cost-efficient way of doing astronomy.

Eclipse Safety

Astronomers, both professionals and amateurs, know that totality of a total solar eclipse is the only safe time that you can stare at the sun with the naked eye. For the partial phases that are seen before and after totality, for the partial phases that are seen by those outside the zone of totality, or for partial or annular eclipses, pinhole cameras or special filters are essential for safe observation of the part of the sun that remains visible (see, for example, the discussion of solar filters in B. Ralph Chou, *Sky and Telescope*, August 1981, pp. 119-121). They cut down the solar intensity until only approximately 0.001 per cent passes. Normally, the eye blink reflex prevents us from looking at the sun for too long. Only in the five or ten minutes before and after totality, or in locations just outside of the zone of totality, are humans usually able to look at the sun long enough to do any eye damage.

Somehow, the astronomers' admonitions that one should never stare at the sun except during a total eclipse, and that one should use proper filters (or pinhole cameras) to see any partial phases, have been converted to widespread public belief that the eclipse itself is dangerous to watch, and even that the sun emits hazardous rays during the eclipse that it does not emit at other times. On the contrary, less of every kind of radiation from the sun is received at an eclipse because the moon hides the rest. The corona becomes visible because the blue sky has faded into darkness, not because the corona becomes brighter.

Over the years, astronomers traveling to eclipses have spent many hours instructing the local population on how to observe the eclipses. Even in many so-called primitive societies, there is widespread interest in a total eclipse, often from oral lore. In the northern frontier district of Kenya, for example, at the time of the 1973 total eclipse, the adults and children of the various tribes were interested in what they would see during totality and asked how to watch it. Children on the Hula peninsula of Papua New Guinea knew just how to watch the partial phases of the 1984 eclipse with pinhole cameras they made (Fig. 16A). Natural pinhole cameras are also eay to find; one can safely watch the shape of the sun projected on the ground (Fig. 16B).

General superstitions in India during the 1980 eclipse kept most people indoors and off the streets. They missed the glory of the eclipse completely. But even in so-called advanced societies, we run into superstitious fears of the eclipse, and failure to comprehend the difference between partial and total phases.

For example, the advance publicity in Manitoba for the 1979 eclipse dealt almost exclusively with how you would be blinded (presumably totally) if you looked at the sun. There was little discussion of why scientists study the eclipse, or even of why the public should look at the eclipse. School boards in Manitoba actually restricted schoolchildren during the entire eclipse to parts of the school away from windows and preferably in rooms without windows, even when the schools were in the zone of totality. One school in Winnipeg even asked for permission to ignore fire alarms if any should go off during the eclipse, lest the students rush outside and be blinded.

The school boards' arguments were based on a letter from the Manitoba Medical Association that absolute safety could not be assured for all students watching of the eclipse. There was no mention that absolute safety is not guaranteed in school buses, in walking to school, in chem lab, in swimming, etc. Indeed, as a result of the school board's action, the absentee rate in Manitoba schools ranged up to 90 per cent. I am sure it was more hazardous to have all those students out of school, walking unprotected (from speeding cars) in the street, than it would have been to have them watching the eclipse from their schoolyards with proper safety instruction.

Students who had been prevented from watching the eclipse surely discovered that they had many friends and relatives who did see the eclipse without being blinded, and that the spectacle was magnificent. The students discovered that other students were bused in from neighboring provinces to see the eclipse, and were unharmed. Many of these students have thus been shown that the school board and doctors cannot necessarily be trusted to tell the truth. How will the students respond if, then, doctors and school tell them that they should protect themselves from AIDS in certain ways? Or not take crack? Won't they suspect the information they are being given is similarly faulty?

The situation was no better in Williamsburg, Virginia, during the 1984 annular eclipse. I could not succeed in reasoning with a principal to allow students in his school to see the eclipse, even though the science teachers supported my position. Fortunately (?), the day was rainy, so the students did not miss anything they would not have missed anyway.

An eclipse can be a motivating experience for students. Some might become so interested in scientific phenomena that they go on to become the scientific leaders and workers of the next generation. Others might simply pay more attention to their schoolwork out of a realization that the universe is so exciting to understand. Instead, the students were bludgeoned with scary advice and given the impression that science and scholarly work is harmful. The consequences are subtle but long-acting. How will these students, when grown up into voters, act when budgets or policies for scientific or medical research and development are put before them?

The ease with which the eye can be harmed is exaggerated in many bulletins. Eclipse eye damage has become rare. Alan MacRobert put the matter in perspective in *Sky and Telescope* (April 1985, page 315), in which he assessed eye damage at the 1984 annular eclipse. He reported that only three cases had turned up. Dr. John C. Cavender, retinologist at the Emory University School of Medicine in Atlanta said that "I know of absolutely no one who suffered retinal damage" other than the three cases. "There have been studies done with humans--people with diseased eyes that were scheduled to be removed--who exposed their retinas to the sun. It's been found that it usually takes minutes to develop solar maculopathy [damage to the central retina]." And even when some damage is found, it often has no effect on vision.

To a certain extent, when an astronomer says, "watch the eclipse safely," we mean it in the same sense in which we may say "drive safely" when sending dinner guests home. We don't mean that something so hazardous is going to happen that you shouldn't watch the eclipse--or that you shouldn't drive home. The hazards of watching eclipses have been so exaggerated in many places that the situation must be brought into better balance for the future.

FUTURE ECLIPSES

The next major total solar eclipse will take place on July 11, 1991. It will cross the "Big Island" of Hawaii, passing directly over the Mauna Kea Observatory, which has no solar telescopes, and unfortunately missing the solar telescopes on Maui. Totality in Hawaii will last 4 minutes 8 seconds. Maximum totality will occur just off the coast of Mexico near Baja California; it will be 6 minutes and 54 seconds there. This long totality will continue over Mexico and over Central and South America. The "saros" effect that eclipses repeat every 18 years 11 1/3 days has long been known; the 1/3 day means that the earth rotates 1/3 of the way around, putting the eclipse in a different location. This eclipse is in the same saros series as the long African eclipse of 1973.

An annular eclipse will cross the United States from southwest to northeast on May 10, 1994.

The next total solar eclipse to cross the continental United States won't occur until 2017. The next eclipse to cross Canada will be in 2024.

ACKNOWLEDGMENTS

I thank the National Science Foundation for supporting several eclipse expeditions and data reduction, most recently on grants PRM-8114631 and RII-8304403, and the National Geographic Society for supporting a series of eclipse expeditions, most recently grant 2547-82. I am grateful to the late Donald H. Menzel, former Director of the Harvard College Observatory, for teaching me about eclipse expeditions. I am pleased to have had the guidance of William Liller, first as a student when he was Chairman of the Astronomy Department of Harvard, and then as a friend.

REMARKS

J. Mattei: Participating in Bill Liller's 1973 solar expedition to Kenya, to study the polarization of the outer corona in multi-color wavelengths, was Bill's wedding present to Mike and me. This is one of the best presents a newlywed couple involved in astronomy could have. Thank you, Bill!

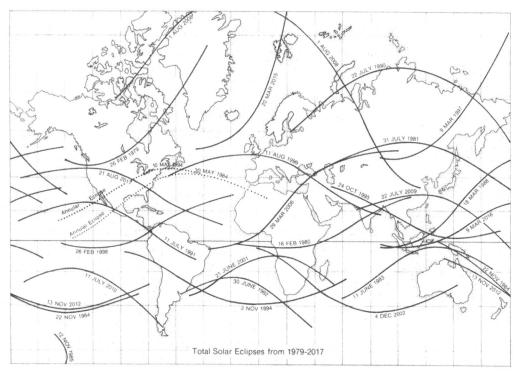

Figure 1. Total solar eclipses from 1979 to 2017, plus the annular eclipses that cross the
United States in 1984 and 1994 (Bryan Brewer, Earth View Inc.).

Figure 2. On the shore of Lake Rudolf (now Lake Tanganyika), at the NSF eclipse site in 1973. (JMP)

Figure 3. The corona at the June 30, 1973, total solar eclipse. (JMP)

Figure 4. Looking through a solar filter made of fogged, developed black-and-white film allows you to look safely at the sun. Here we see the view during annularity at the 1973 annular eclipse. (JMP)

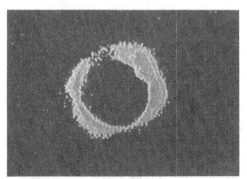

Figure 5. The solar corona in the iron-XIII line at the 1977 total solar eclipse, viewed from a ship in mid-Pacific. (Maxwell T. Sandford of Los Alamos Scientific Laboratory and JMP)

Figure 6. The corona in the sky at the 1979 eclipse in Brandon, Manitoba. (JMP)

Figure 7. Totality in India during the 1980 eclipse. (JMP)

Figure 8. The diamond ring effect, enlarged because of internal reflections in the airplane window, and the corona seen from an airplane off Kauai during the 1981 eclipse. The light reflected in the ocean below comes from the sun outside the zone of totality. (JMP)

Figure 9. Totality at the NSF site at Tanjung Kodak, East Java, during the 1983 eclipse, looking into our heliostat. An image of the sun with our optical fibers and their holders silhouetted is at lower right. (JMP)

Figure 10. Totality at the 1984 eclipse in Papua New Guinea, with the sun 23° in altitude. (JMP)

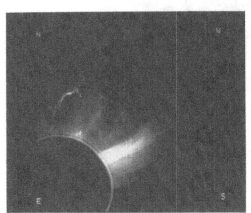

Figure 11. The occulting disk of the Coronagraph/Polarimeter aboard the Solar Maximum Mission is 1.75 solar radii in radius, and is surrounded by diffraction rings. (HAO/NCAR/NSF and NASA)

Figure 12. The corona in the coronal green line with the Haleakala coronagraph. (Institute for Astronomy, University of Hawaii)

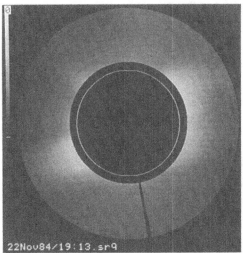

Figure 13. The corona, from polarization measurements with the Manua Loa coronagraph. (Courtesy of D. Sime, HAO/NCAR/NSF)

Figure 14. Coronal structure from calibrated photographs taken at the 1984 eclipse in Papua New Guinea, to which we have applied a radial-filter algorithm. (JMP and John MacKenty, Space Telescope Institute)

Figure 15. The repaired Solar Maximum
Mission, about to be returned to Earth orbit.
(NASA)

Figure 16A. Children in Papua New Guinea
with their pinhole cameras, watching the partial
phases of the 1984 eclipse. (JMP)

Figure 16B. Interstices among leaves act as
pinhole cameras for these children watching the
1984 annular eclipse.

Galactic Astronomy

PLANETARY NEBULAE, NEW INSIGHTS AND OPPORTUNITIES

Lawrence H. Aller
Astronomy Department
University of California
Los Angeles, CA 90024 USA

1 SOME RECOLLECTIONS

One of the greatest good fortunes of this scientist's life was not only to have met Bill Liller but to have been able to entice him to work on the photometry and spectrophotometry of gaseous nebulae, in particular, planetary nebulae (PNs). At this epoch in the early 1950's the study of what we now call the diagnostics of nebular plasmas was severely hampered by the lack of good observational data. Photographic spectrophotometry of PNs had been initiated by Louis Berman in 1927. Essentially the same methods were still in use at the end of the 1940's, although improvements in stellar spectrographs, the aluminization of telescope mirrors, and a better understanding of the foibles of photographic plates led to gradually improving results. Nevertheless, the nonlinearity of photographic emulsions combined with the great range in nebular line intensities led to severe difficulties.

Bill Liller undertook to do something about this situation. He mounted a photocell at the focus of the 0.6-m Curtis Schmidt telescope over whose aperture an objective prism had been placed. In spite of the faintness of PNs, none of which are visible to the naked eye, it was possible to measure the green nebular lines and Hβ. At this epoch one still had to rely on photographic plates to get the 4363 [O III]/4340 Hγ ratio in order to obtain electron temperatures, T$_\varepsilon$ (Liller and Aller 1954). MacRae and Stock (1954) had made similar measurements of a few planetaries at about the same time. In 1956, Liller built a photoelectric spectrum scanner with which he and colleagues made measurements of gaseous nebulae, stars, and comets. In that year spectral scans were secured with the 60- and 100-inch telescopes at Mt. Wilson to cover both ordinary "optical" and near-infrared (IR) regions (Liller and Aller 1963). This Liller scanner was one of the very first to see service in the southern hemisphere where it was used to measure stars, H II regions, PNs, and globular clusters. Similar devices subsequently have been used successfully on PNs by O'Dell, Osterbrock, the Peimberts, their respective colleagues, and many others.

2 STELLAR EVOLUTION: THE PLANETARY NEBULAE MOVE TO CENTER STAGE

In this review, planetary nebulae are emphasized, a topic on which much of Bill Liller's early research was concentrated. The present point of view is very different from that of 35 or 40 years ago. Then the concern was primarily with the "nitty gritty" of acquiring fundamental diagnostics, T_ε, densities, N_ε, chemical compositions, and gross structural features for these objects. Distance determinations were, and remain, an elusive goal (see §8). The complete picture of stellar evolution was yet to emerge; it developed that PNs were destined to play a critical role in this drama. They were the envelopes cast off by dying stars as they neared the end of their nuclei-burning lives and headed for oblivion as faint, dead white dwarfs (WD's).

As a star ascends the giant branch for the second and last time, the highly distended envelope is detached and ejected into space. This low-density shell may form a PN as the core settles down to a hot sdO star and ultimately a WD. In the scenario by Kwok *et al.* (1978) a fast wind from the hot core bulldozes the sluggishly moving former envelope of the defunct giant star into a shell of enhanced density, which is now exposed to the rich ultraviolet (UV) radiation flux of the sdO core. Variations on, and modifications of, this scheme have been proposed by a number of writers. In any event, the subsequent development of the PN is largely influenced by what happens to the hot core that has now become the PN nucleus (PNN).

3   EVOLUTION OF A PLANETARY NEBULA NUCLEUS (PNN)

At the end of the asymptotic giant branch (AGB) evolution (i.e., the ascent of the giant branch for the last time) the star loses most of the outer envelope – stripping off most of the material above the carbon/oxygen (CO) core. Details of the process are obscure. The star may become a Mira or an OH-IR variable during this phase of its evolution, which has been considered by Paczyński (1971), Schönberner (1981, 1983), and Iben (1984).

When the mass of the envelope has been reduced to about 0.001 that of the sun, the surface temperature rises rapidly. The H and He remaining in the thin shell overlying the core may be burned. H is presumably burned quiescently, while He is consumed during shell flashes. The course of subsequent PNN evolution and nebular development depends on whether nebular ejection occurs during the H-burning or He flash phases. Not all of the envelope may be lost at the time the material destined to form the PN is ejected. Iben (1984) suggests that about one-half of all PNNs are burning He rather than H. Wood and Faulkner (1986) argue that every AGB star that can produce a core more massive than 0.86 $M_\odot$ will result in a PN with a He-rich PNN (unless it is so massive as to blow up as a supernova!). They find that for the vast majority of PNNs which are less massive and that leave the AGB during the interflash stage, the transition time from AGB stage to PN stage depends strongly on the phase of the shell flash cycle at which the star leaves the AGB.

The star heats up to a surface temperature of about 10^5 K and its luminosity remains constant for a time. In the ordinary run of PNN evolution, the whole of the remaining envelope has now been burned into He or heavier elements. A PNN powered by He burning evolves relatively slowly, but an H-burning PNN may fade by ten-fold in about 100 years.

The evolutionary time scale depends critically on the mass, M, for a core of about 0.6 $M_\odot$. If the PNN envelope is being blown away in a wind on a time scale comparable with its nuclear burning rate, the pace of PNN fading may be considerably greater. Schönberner's H-burning stars may be the most frequent type of PNN. Since the intrinsically most luminous PNNs will decline most rapidly, the fading of these stars should be readily observable. Liller's pioneering photoelectric measurements of a number of these stars, and work by Liller and Shao (1970), and by Kaler and his associates, may have already provided a baseline sufficient for an early test of this hypothesis. The lifetime of the fragile PN is short compared with the aeons required for the CO sdO core to decline from the status of a PNN to final crystallization. Iben and Tutukov (1985) have traced the rather complicated effects of neutrino losses, liquefaction, and finally crystallization, to the point where the WD cools to earth-like temperatures.

Since the evolutionary rate depends so strongly on the mass, it behooves us to obtain as accurate masses as possible for PNNs. Schönberner (1981) argued that nearly all PNNs fall in a narrow range 0.55 $M_\odot$ < M(PNN) < 0.64 $M_\odot$, although Kaler suggested that 30% might have masses greater than 0.64 $M_\odot$. Heap and Augensen (1987) suggested that moderately massive PNNs would be found in young, optically thick nebulae, but the most massive objects would appear as white dwarf-like objects in optically thin nebulae. They concluded that if M(PNN) < 0.55 $M_\odot$, the nebular shell would escape before the star heated up enough to excite it. Between 0.55 and 0.64 $M_\odot$, the PNN remains bright over the lifetime of the nebula. Objects between 0.64 and 0.75 $M_\odot$ could be found in young, large nebulae before they faded, but those with masses exceeding 0.75 $M_\odot$ would dim quickly. The lower mass, 0.55 $M_\odot$, PNNs come from objects near a solar mass but those with M(PNN) > 0.64 $M_\odot$ would correspond to progenitor stars with masses greater than 1.3 $M_\odot$. On the other hand, Grewing, Bianchi, and Cerrato (1986) find M(PNN) > 0.95 $M_\odot$ for NGC 40 and M(PNN) > 0.78 $M_\odot$ for NGC 6543. These stars should be fading quickly; time-dependent effects should appear in the PNs.

A frequent procedure for obtaining M(PNN) is to assume some distance scale, choose temperatures from the Zanstra method, and then use theoretical evolutionary tracks to acquire the visual brightness, L_V(PNN) as a function of time, choosing a proper M(PNN) to fit observations to theory. Weaknesses in this method are the assumption of a distance scale, and an often uncritical choice of Zanstra temperature, T(Z). For O- and Of-type stars that show absorption lines, an alternative procedure is to try to retrieve T(*) and log g(*) from the spectrum. Theoretical evolutionary tracks give a T(*) versus log g(*) relationship, so we can infer M(*) and R(*) directly from them and obtain an independent distance estimate as well. An early attempt to apply spectral data to procure T(*) and M(*) (Aller 1948) was frustrated by severe limitations imposed by photographic photometry and the relatively small spectral dispersions then available. Fortunately, Méndez et al. (1988) were able to obtain high-resolution spectra of O and Of stars with a charged coupled device (CCD) detector. They were able to calculate effective temperatures, log g, and He/H ratios with the aid of a non-LTE model atmosphere. The emission-line O stars seem to be slightly more massive than absorption-line O stars. They lie closer to the Eddington limit. Of the sample of 21 PNNs which appear to have evolved from AGB stars, nearly one-half have masses exceeding 0.70 $M_\odot$, and the mean mass of the group, 0.69 $M_\odot$, exceeds the mean mass found in other studies of PNNs. Alas, the nebular ages deduced from size and expansion rates do not agree well with theoretical predictions. In summary, while most PNNs lie in the range 0.55 to 0.64 $M_\odot$, the O and Of stars studied by Méndez seem to constitute a sample consisting of more massive objects. One possible word of caution: the theoretical model atmospheres are assumed to be in hydrostatic equilibrium. In view of the widespread phenomenon of winds, how good is this assumption?

How hot can the nuclei of PNs actually become? Many estimates have been made, mostly by Zanstra-type methods (see e.g., Pottasch 1984; Kaler 1983; Shaw and Kaler 1985; Reay *et al.* 1984). The Zanstra methods normally employ only He I and He II lines. One alternative procedure is to use the entire level of excitation of the spectrum rather than simply the helium lines to get a leverage on the UV flux. That is, we use nebular models in which the stellar flux incident on the nebula is an adjustable parameter. For example, the spectrum of NGC 6741 would be reproduced with a Planckian distribution at T = 1.6×10^5 K, but with a depression in the far-UV to avoid too strong [Ne V] fluxes. For the very high excitation planetary, M1-1 (Aller, Keyes, and Feibelman 1986), even a 2.25×10^5 K black body flux would not work. Finally, a black body at 1.8×10^5 K was used, but with the far-UV (7.2-9.0 Rydbergs) flux enhanced by a factor of 2. Kaler and Feibelman (1985) measured the ultraviolet spectra of the nuclei of large PNs. Generally, the fluxes fitted the atmospheric models of Hummer and Mihalas or black bodies with temperatures at 10^5 K or above, but some slopes were steeper than Rayleigh-Jeans curves. They measured color temperatures by comparing UV and optical region fluxes. Curiously, the steepest slopes did not come from the hottest stars but very hot, high-luminosity stars had high, determinable color temperatures, so the color temperatures may represent an upper limit to the real temperatures. Theoretical fluxes seem to work at the lower temperatures but not at the very highest. For some objects such as the nucleus of NGC 2440, Pottasch and his associates have proposed Zanstra temperatures exceeding 3×10^5 K which seem to be surprisingly high (see §6).

What about mass loss? Winds appear to be present in all low-excitation, young PNNs but subside as one moves on to more (aged?) compact objects (Cerruti-Sola and Perinotto 1985). At higher spectral dispersion, detailed line shapes have been measured and analyzed by the theory of radiatively driven winds (Castor *et al.* 1975; Abbot 1982). Bombeck, Köppen, and Bastian (1986) found that even in the densest winds, atomic levels were all populated by radiation rather than by collisions. The ionization level in the wind is strongly correlated with the excitation level of the nebula; stellar photons not only determine conditions in the PN, but also set the ionization of the wind. The mass loss rates are difficult to assess;

values of 10^{-8} to 10^{-10} $M_\odot$/year are typically quoted. For the central star of NGC 3242, Hamann et al. (1984) suggest that the mass loss per unit surface area is comparable with that from the Population type I highly luminous ζ Puppis. Bombeck et al. (1986) argue for mass loss rates as high as 10^{-7} $M_\odot$/year, but Heap (1986) finds terminal velocities in PNNs to be smaller than in Population I stars. She notes that the terminal velocities of 600-3600 km/sec are strongly correlated with stellar temperature, and concluded that all PNNs must have closely comparable masses. As for the immediate post-AGB evolutionary stage, following the possible evolution of the star as a Mira or OH-IR object, perhaps mass loss rates as high as 10^{-5} $M_\odot$/year occur.

Several PNNs are binaries. Among the best studied is the nucleus of Abell 41, a He-rich sdO star (Green et al. 1984; Grauer and Bond 1983). With a grid of very high gravity, local thermodynamic equilibrium (LTE) atmospheres, they found $\log g = 6.0$, $T = 40,000$ K, M(star) = 0.6 $M_\odot$, R = 0.13 $R_\odot$. The double-lined spectroscopic binary LSS 2018 studied by Drilling (1985) has P = 8.57 hrs. There are no eclipses, so it is possible to deduce only limits of the inclination $53° < i < 72°$, 0.4 $M_\odot$ < M(primary) < 0.7 $M_\odot$, and 0.2 $M_\odot$ < M(secondary) < 0.3 $M_\odot$, which fit Schönberner's theory. Feibelman and Kaler (1983) found the true nucleus of the giant PN (339+881) to be a close binary companion of a G5 star, SAO 82570; it is one of the hottest known PNNs with a luminosity perhaps 600 times that of the sun.

The low-amplitude ($\sim$ 0.02 mag), pulsating PNN of Kohoutek 1-16, studied by Grauer and Bond (1984) has a main period of about 28.3 minutes but several other periods seem to be present. This star may be related to the non-radially pulsating ZZ Ceti stars which are WD's but its high temperature excludes membership in this class.

4   THE PLANETARY NEBULAR EVENT

Although the ram pressure of the central star wind appears to be the dominant phenomenon in the hydrodynamical evolution of a PN, other factors may have to be considered: thermal expansion, radiation pressure on dust grains, and shock waves. At the epoch of ejection there also may appear a "superwind" (Renzini 1981) that can carry substantial momentum and energy away from the star.

In the Kwok model (Kwok et al. 1978; Volk and Kwok 1985) the wind which blew from the AGB star just before the PN event provided most of the mass, while the wind from the uncovered core supplied the energy and momentum observed in the shell expansion. They attribute the higher density, well-defined shell structure of PNs to an isothermal shock that separates shell and red giant wind. The fast wind impinging upon the slowly moving cloud ejected by the red giant causes a shock that is reflected backwards towards the PNN and into the low-density region. This shock raises T_ε to > 10^6 K. X-ray emission from this cavity would be weak and probably undetectable. Volk and Kwok considered that because of the small cooling rates involved, the energy in the wind is conveyed to the shell with efficiency high enough to account for the observed expansion of the nebular shell so that "PNs can be formed by the interacting wind process alone."

In its initial stages the PN cannot yet be observed at all, except perhaps as an IR object. We refer to this epoch as the "proto-PN" stage. Only when the stellar temperature reaches a value of about 30,000 K does the excitation in the nebular shell rise sufficiently for it to be observed. Volk and Kwok argue that this proto-PN stage cannot last much longer than 1500 years, and during this time the PNN evolves rapidly. The PNN may emit sufficient photons to ionize and excite the nebula over only a few thousand years and the PNN wind may fluctuate over this time. Thus, theoretical studies of PN evolution must take into account the simultaneous evolution of the PNN.

Sabbadin *et al.* (1984) propose a somewhat different scenario. The red giant approaching the end of its AGB life ejects material at a low velocity. When the outer envelope becomes uncoupled from the interior and the star begins its rapid march to a state of higher surface temperature, in a short interval of time it ejects a substantial mass of gas with a range of velocities. This ejected blob or shell interacts with the previously expelled material of the stellar wind. In the early phases we do not see the interaction region between the gas blob and the cool, already established, slowly moving wind from the AGB star. We see only the inner Strömgren sphere which slowly advances into the cool gas. As the nebular shell expands and its density falls, the edge of the Strömgren sphere finally reaches the interaction radius that marks the boundary between the blob ejection and the previous AGB wind. Thenceforth, the observed effects are governed by the interaction between the ejected blob or shell and the low-density AGB material. In this model the AGB gas is compressed slightly, if at all, and the density in the outer region is so low that the emitted radiation can scarcely be detected. Sabbadin *et al.* (1984; Sabbadin 1984, 1986) have observed radial velocity fields, structures, and expansion rates for several PNs and have concluded that the data support their hypothesis.

Other models have been calculated in an attempt to take into account the concurrent evolution of PNs and PNNs. For example, Vilkoviskii *et al.* (1984) adopted Schönberner's (1981) sequence of PNN models to calculate a series of static models with which they were able to explain some features of nebular line intensities as a function of nebular size. Okorokov *et al.* (1985) considered a dusty variable mass loss model and Paczyński's (1971) evolutionary sequence.

The most elaborate investigation of the influence of PNN evolution on PN development is that given by Schmidt-Voigt and Köppen (1987) who also employed the Schönberner evolutionary tracks. At an early stage the star's temperature rises while its luminosity remains fixed so the supply of UV photons increases. This is the "ionization phase." The character of the PN's fate depends on the mass of the central star and its rate of mass loss. For a PN of about 0.25 $M_\odot$ excited by a 0.6 $M_\odot$ star, this phase lasts about 6000 years. Later, as the central star's luminosity falls rapidly, the nebula recombines and the nature of the old AGB wind becomes important. These authors prefer a three-wind model to Kwok's two-wind model. It consists of (a) the wind from the AGB precursor star, (b) Renzini's superwind losing about 10^{-4} M$_\odot$/year for 1000 years, and (c) Kwok's fast wind from the central star. They conclude that the existing PN contains a strong component of material from the wind of the now-defunct AGB star. In the final phase, at least one-half of the material comes from the old red giant wind, thus implying a large mass loss rate in the progenitor star. In their scenario, the formation of a PN includes the injection of material from a superwind that lasts about 1000 years. In particular, a PN involves much material from the AGB wind phase, thus requiring a brisk wind during this epoch.

These models suggest a number of interesting observational tests; it is by no means clear that every PN is formed by the same routine. It is possible that in some objects we may have a situation resembling a superwind, in other situations not. In the Schmidt-Voigt–Köppen and Sabbadin scenarios we might often expect pronounced abundance variations since the chemical composition of the AGB wind may differ from that of underlying stellar layers. In some PNs such as NGC 7662 (Barker 1986), there appear to be no large composition differences from point to point. In the famous case of A30 there exist large abundance differences between some interior knots and outer regions. For the vast majority of PNs, low spatial resolution confines our data to a single bright rim or shell or to the integrated light of the entire PN.

Dramatic differences between the chemical compositions of PNNs and surrounding PNs have long been known to exist, especially for a number of instances in which carbon-rich WC stars excite PNs of nearly normal composition. An early example that comes to mind is Campbell's "hydrogen envelope star," BD 30°3639. This nebula shows strong [N II] lines, yet there is no N in the PNN spectrum. Ingenious mechanisms for selectively expelling N by radiation pressure were proposed but then O (prominent in the atmosphere of the PNN) should have been dragged out as well. A similar, dense young PN is SwST1. The nucleus is a WC10 carbon star while the envelope is O-rich with C/O = 0.7 and the IR spectrum displays silicate features.

5   NEBULAR STRUCTURES AND SHAPES

Progress has been made in interpreting the forms displayed by PNs. We see only the two-dimensional projection of three-dimensional structures but by using high-resolution velocity and spatial data it seems possible to construct realistic portrayals of many PNs. Much effort has been spent in describing the often complex forms exhibited by these nebulae. Evidently most of them can be understood in terms of the fairly simple precepts which have been given by Bruce Balick (1987). The simplest situation is that of a spherically symmetrical shell of AGB material which is suddenly subjected to the wind from a hot bubble created by the wind from the uncovered core. Since the velocity of the wind, typically 800 to 3000 km/sec, exceeds the speed of sound in the gas, it creates a shock front that compresses the gas ahead of it. The advancing front eventually breaks through the shell and fragments it into clumps and filaments. Blobs and clumps of gas can occur in seemingly spherically symmetrical structures. The radiation from the star ionizes the gas and different situations prevail for ionization-bounded versus material-bounded objects. In the course of its evolution a nebula may evolve from an ionization-bounded to a material-bounded stage.

The original AGB shell is often not spherically symmetrical; the density and thickness of the shell may depend on latitude. That is, an evolving, rotating star may eject a much larger quantity of gas in its equatorial plane than in the polar direction. The scenario is now quite different. The expanding bubble of hot gas encounters a shell that is thick near the equator and thins out towards the pole. The nebula becomes elongated in shape and finally the bubble will burst through at the poles. Furthermore, in the equatorial plane the PN is very thick optically so the Strömgren sphere limit lies well inside the material boundary of the nebula. The entire polar region may be optically thin. Combining the effects of the hot wind that blasts a hole through the polar regions with the optical depth effects, we may easily see how PNs with symmetrical lobes may be produced. NGC 7009 is a classic example. By viewing such nebulae from different angles, a variety of structures can be envisaged and when one recalls that the ejected shells may have pronounced density fluctuations, condensations, and so forth, a gamut of very complex forms can be anticipated.

From a study of the morphology of 108 nonstellar planetary nebulae, Zuckerman and Aller (1986) found 50% to show a bipolar symmetry and 30% to show elliptical symmetry. There was no indication that bipolarity is a characteristic peculiar to carbon-rich evolved stars. In other words, whether a PN evolved from an AGB precursor with an enhanced equatorial ring or from a spherically symmetrical envelope seemed to have nothing to do with processes that would give a greater production of C by the triple-α process. On the other hand, N-rich objects such as NGC 6302, NGC 2440, and others seem to be primarily bipolar in structure (Peimbert and Torres-Peimbert 1983).

Balick and his associates (1987) made detailed studies of (a) the bright, morphologically complex PN, NGC 6543, (b) one of the most spherically symmetrical objects known, IC 3568, and (c) a filamentary, low-excitation object, NGC 40. Both NGC 6543 and IC 3568 show large density changes on scales corresponding to sound crossing times less than one century. Possibly the density gradients are maintained by wind-driven shocks. The kinematics and morphology of NGC 6543 are peculiar and difficult to understand. NGC 40 is excited by a WC star. The [O III] emission which is much weaker than in most planetaries is produced in a smooth shell that is interpreted as the interface between the hot (10^5 to 10^6 K) interior and the cool (8000 K) exterior. The filamentary nature of the outer shell may reflect the action of shock waves in the ionized remnant of the AGB shell. These results by Balick and his associates demonstrate the importance of very detailed structural and kinematical studies of individual objects.

Models of other PNs have been constructed by various workers. One of the most intensively investigated objects is the so-called "Eskimo Nebula," NGC 2392. The inner ring appears to reveal a prolate spheroidal shell which we see almost pole-on, while the outer ring corresponds to a more nearly spherical shell. The geometrical models proposed by Reay *et al.* (1983) and by O'Dell and Ball (1985) are in rather good accord, but the velocity patterns found by the two teams differ substantially. The bipolar PN, NGC 7026, shows a pronounced ionization stratification and dependence of expansion velocity on different zones and ion types. Solf and Weinberger (1984) suggest that the equatorial plane of the toroid is inclined by about 15° to the line of sight. The low excitation, [N II] and [S II] zone of the ring and the polar blob expand at comparable rates. The kinematical age is about 800 years.

To what extent does the binarity of PNNs play a role? A possible influence of the presence of a stellar companion upon nebular morphology was considered by Livio, Salzman, and Shaviv (1979), Morris (1981), and Livio (1982). From a comparison of the sizes of extreme red giants and the separations of solar-type, main-sequence double stars, Zuckerman and Aller (1986) concluded that the gravitational effects of companions are not likely to produce mass ejections, at least not in most cases. It is quite possible that the presence of a companion star in a PN progenitor system can influence the symmetry of the evolving nebula. The number of PNNs occurring in known binary systems does not suffice for satisfactory statistics.

6   *VARIABILITY IN PLANETARY NEBULAE*

In 1937, while I was working as a summer assistant at Lick Observatory, Louis Berman asked me to observe IC 4997 because he suspected that the 4363 [O III]/4340 Hγ ratio had changed since the early part of the century. In 1938, I found the ratio to be 1.6 by photographic photometry. In 1956, the first measurements with the Liller spectrum scanner showed, much to the surprise of everyone, that this ratio had dropped to 0.76. It continued to fall and then seemed to stabilize around 0.62 in the early 1970's. Feibelman *et al.* (1979) found rapid variations: 1.0 (1977.7), 0.91 (1978.3), and 0.8 (1978.6). This latter rise was interpreted as an indication of a rise in electron temperature, perhaps a result of some transient event. Because of its high density ($N > 10^5$ cm^{-3}), and therefore quick response time, this PN is sensitive to rapid flux variations in the PNN energy output. These would not be observed in low-density nebulae with long time constants. The most dramatic example of the variation of a PNN is, of course, FG Sagittae (Langer, Kraft, and Anderson 1974). Here the response of the nebula would be very slow. Harrington and Marionni (1976) have discussed this problem for the PN associated with FG Sag.

One of the most engaging results is the apparent association of a subset of R Coronae Borealis stars with PNs. Iben *et al.* (1983) suggested than an R Cor Bor star may be produced when the nucleus suffers a final helium shell flash. The PNN then bloats to a giant star with temperature, luminosity, and chemical composition appropriate to an R Cor Bor star. Schaefer (1986) has identified several R Cor Bor stars as PNNs; they are much too cool to excite the presently observed nebulae.

One of the most remarkable objects is NGC 2346 (see Schaefer 1985 for references). The observational record showed that from 1899 to 1981, at least, this PNN had not varied. Then, suddenly, large amplitude eclipses appeared, the light curve of the PNN changed rapidly, and these strange eclipses increased in duration until they covered most of the orbital phase. Schaefer suggests that this unique behavior is caused by occultation by a dust cloud which had condensed from a shell ejected by the PNN. The eclipses appear to have ceased; evidently, the cloud has dissipated. Another R Cor Bor-type PNN appears to be V348 Sgr, studied by Dahari and Osterbrock (1984). They concluded that the original H-rich envelope had been expelled and a final He shell flash in the outer layers of the remnant core formed a carbon-rich, extended atmosphere.

Are time-dependent effects easily identified in ordinary PNs? Since recombination times are of the order of $10^5/N_\varepsilon$ years, we might expect that for a PN with $N_\varepsilon \sim 10^3 - 10^4$ cm^{-3}, such effects could be seen if the star suddenly suffered a drop in brightness. Theoretical models of PNNs which burn H quiescently predict a rapid luminosity fall when the star finally leaves the horizontal branch and plummets as a white dwarf. Tylenda (1986) has investigated this situation and suggests that the sudden luminosity decrease could cause a double-ring structure to appear in PNs. There will be an inner, bright, high-excitation ring that is surrounded by a faint, low-excitation halo that fades with time as the plasma recombines. On the other hand, Hippelein et al. (1985) argue that dim haloes with surface brightness in the range of 10^{-3} to 10^{-4} that of the main nebulae favor Sabbadin's (1984) theory. That is, at this late evolutionary stage, the inner part of the PN which had been produced by the massive shell ejection becomes optically thin and the UV photons now excite the low density remnants of the attenuated AGB wind.

Some notable discrepancies in PN theory may be attributable to evolutionary time-dependent effects. For NGC 2440, Walton et al. (1986) find a Zanstra temperature of 3.5×10^5 K. Their measurement of the magnitude of the faint PNN of this high-surface-brightness PN was a technological triumph combining skill, narrow-band-pass filters, and good luck with the seeing. Since all the optical-region radiation of a PN is derived from the degradation of UV photons, a magnitude difference $m(*) - m(\text{neb}) = 18.9 - 8.9 = 10.0$, requires a huge T(Z) if the nebula is in equilibrium as required for applying the Zanstra theory. A temperature as high as 3.5×10^5 K apparently cannot be reconciled with the excitation level of the spectrum (Shields et al. 1981). That is, if we calculate a theoretical nebular model and fit computed to observed line intensities, we find that $T(*) = 1.7 \times 10^5$ K. A similar effect is found in NGC 6565 for which Reay et al. (1984) find $m_V(\text{PNN}) = 19.2$ and deduce $T(Z) = 1.5 \times 10^5$ K, at which temperature [Ne V] would be very prominent. Is it possible that the PNNs are fading or variable? If so, we should soon see large effects in the excitation of the nebular spectra. In that case, we would have to solve a time-dependent nebular model problem as has been treated by Harrington and Marionni (1976) and by Tylenda (1986).

7   *DUST, COOL GAS, AND MOLECULES*

In the optical regions of the spectrum, the evidence of dust and molecules is subdued. Infrared observations through windows in the earth's atmosphere, from the Kuiper Airborne Observatory and the Infrared Astronomical Satellite (IRAS), have opened new vistas in this important spectral region. In the near-IR, shortward of about 2 μm, the continuous spectrum is dominated by free-free emission. Farther in the IR, the thermal emission from dust becomes much more important.

In most PNs the dust and ionized gas appear to be well mixed, as in NGC 7027 (Becklin *et al.* 1973); Basart and Daub (1984) found the temperature distribution in NGC 7027 to resemble that in the optical image, suggesting that dust is heavily concentrated in H II regions and is involved in the energy balance of the gas. In IC 3568, Cohen *et al.* (1984) find the far-IR excess to be much greater than would be expected from the 10 μm data. They conclude that here dust coexists with the ionized gas and is heated by the stellar flux. The dust grains appear to resemble amorphous carbon. In some objects such as IC 418, much of the dust emission may be concentrated outside the H II region. Bentley *et al.* (1984) find that the material producing the diffuse lines at 8.6 and 11.3 μm seems to lie at the edge of the nebula. The composition of the dust seems to be sooty in C-rich objects. In O-rich envelopes, it consists of silicates.

In outer, cool envelopes, material of the original AGB wind exists largely in molecular form (Mufson *et al.* 1975; Zuckerman *et al.* 1976). A search for 21-cm emission by Rodriguez and Garcia-Barreto (1984) with the VLA revealed atomic H only in NGC 6302, while Glassgold and Huggins (1983) suggest, on theoretical grounds, that detectable H I may exist only in PNNs with progenitors with surface temperatures above 2500 K and in systems with strong N enhancements. Isaacman (1984) finds molecular hydrogen in J900, NGC 2440, NGC 7662, IC 5117, M1-78, and possibly in NGC 6210, but not in NGC 6572, NGC 6884, IC 2149, or IC 2165. The strengths of the observed transitions cannot be explained by radiative excitation. Kwan's (1977) shock model looks more promising. It predicts excitation temperatures from 1000 to 4000 K behind the shock fronts with shock velocities of 10 to 24 km/sec. Clearly, the molecular hydrogen in NGC 7027 is excited from outside the ionization front. Since such a collisionally excited gas is in thermal equilibrium, as radiative events are relatively rare, one can obtain the excitation temperature immediately. It turns out to be about 1200 K. One can also use the shock model to obtain the pre-shock densities in the gas and these are similar to the electron density in the ionized region. The molecular hydrogen mass in NGC 7027 amounts to about three solar masses (Beckwith *et al.* 1980). A similar large mass is found for M1-78. Such large masses imply a very massive progenitor star; NGC 7027 is clearly a far-from-typical PN.

One of the most surprising results is the large molecular hydrogen envelopes found in the Dumbbell Nebula NGC 6853 and in NGC 2346. These extend far beyond the visible boundaries of the nebulae. The binary PNN in NGC 2346 is still girdled by a dense molecular cloud. These types of measurements will be of crucial value in establishing the total masses of the ejected shells since now the contributions of both neutral and ionized gas will be available to us. For nebulae only optically observed, it is of course possible to get only the mass of the ionized portion of the nebula, provided the distance of the object is known (see Zuckerman 1987).

8   THE ABIDING PROBLEM OF NEBULAR DISTANCES

The "bone in the throat" of the PN researcher is the determination of reliable distances for individual objects. We know the answers for PNs in the Magellanic Clouds but structural information is difficult to obtain for these nebulae, although Barlow *et al.* (1986) have been able to resolve SMC N2 by speckle interferometry and to show that it has a double ring structure, the shells having radii of 0.06 and 0.10 pc. The total mass of the PN is 0.36 $M_\odot$. Wood *et al.* (1986) have measured diameters of 11 Magellanic Cloud PNs by speckle interferometry. These objects are small (< 0.13 pc), dense, and young (age $\lesssim 1500$ years), and partially ionized with 0.006 $M_\odot \lesssim M(H^+) \lesssim 0.19\ M_\odot$.

Planetaries with binary PNNs often yield trustworthy distance data but these systems are very few in number. For the average field PN we have recourse to only statistical methods, such as τ and v components of proper motion and radial velocities.

The method of using angular diameters, some assumption about the nebular structure, the Hβ flux, and an assumed mass, was first used by Minkowski and Aller (1954) to set constraints on the distance of the Owl Nebula, NGC 3587. Since the mass of the ionized shell was neither known, nor deemed likely to be constant from object to object, this procedure for distance determination was not considered promising by these authors. Shklovsky (1956) argued, however, that the mass of ionized gas was sufficiently constant to justify the use of this method, which has often been used to determine PN distances. Recent work (see, e.g., Gathier *et al.* [1983] who find the ionized gas mass to vary by a factor of more than 40 for some 42 PNs near the galactic center) emphasizes that the Shklovsky method must be used with caution.

Kwok (1985) has raised a very serious and fundamental objection to the use of radio frequency (or Hβ) and angular diameter data for distance determinations. He considered the evolutionary models of Schönberner (1983) for $M(\text{PNN}) = 0.6\ M_\odot$ and found that the predicted rapid evolution of the central stars implied that the evolution of the PN was thereby strongly affected. The dependence of the radio-frequency or Hβ fluxes on angular size as a result of the expansion of the ionization front had nearly the same mathematical form as the flux-angular size dependence for PN at different distances. Thus, the distance of an ionization-bounded PN cannot be found from the flux versus angular size data alone.

Methods involving the H I 21-cm absorption and reddening of nearby field stars have been proposed. Recent examples are Gathier, Pottasch, and Pei (1986) and

Gathier, Pottasch, and Goss (1986). For a few objects it may be possible to apply a method proposed by Kaler *et al.* (1985). If the central star has a strong wind that gives a P Cygni profile, we infer the escape velocity from the terminal velocity and thus obtain a functional relation connecting stellar temperature, luminosity, and mass. The theory of stellar evolution gives a second relation between these quantities so if the PNN temperature is known by the Zanstra method, we can acquire the stellar mass and luminosity and hence the distance, since the apparent stellar magnitude and interstellar extinction are known. The method seems to be a reasonable one, although it is applicable to only a few objects.

The distances of objects such as novae and the Crab Nebula have been measured by the method of expanding shells. For applications to PNs see Liller and Liller (1968). An intrinsic difficulty with PNs is that we may be measuring only the expansion of the ionized zone into the neutral shell perpendicular to the line of sight. If the PN has a number of distinct blobs and condensations, and a suitable three-dimensional model can be constructed for it, we may be able to determine the distance by measuring both the "proper motions" and the radial velocities of individual blobs. Terzian (1987) has suggested that by using the VLA, we can measure true angular expansion rates together with radial expansion rates in km/sec. This method might prove practical only for symmetrical objects. It may well be that accurate distances for individual, arbitrarily selected PNs will turn out to be one of the greatest technological achievements of PN research!

As mentioned earlier, for O-type PNNs showing absorption lines, it is sometimes possible to obtain effective temperatures and log g from which Méndez *et al.* (1988) were able to establish individual PN distances. These proved to be larger than those obtained by older methods (e.g., Daub 1982; Pottasch 1984), but when this method was applied to a hot subdwarf in the globular cluster NGC 6397, it resulted in the same distance measurement to the cluster as did other methods.

Curiously, although it is difficult to get distances of individual PNs, these objects have been used in an independent method for getting distances of galaxies, e.g., M81. Why? Studies of planetaries in the Magellanic Clouds and other nearby galaxies show that there is a well-defined upper limit to their luminosities, an absolute visual magnitude of −4.0 (Jacoby *et al.* 1987a). This result seems to support the Méndez distance scale.

9   *SOME FURTHER OPPORTUNITIES*
The Magellanic Cloud planetaries offer some big challenges. Here are stellar populations at a known distance; therefore, the problem of distance is traded for one in spatial resolution. All these PNs appear to have sizes less than about $1''$.

From an investigation of expansion rates and other statistics of Magellanic Cloud PNs, Dopita *et al.* (1987) found that the ionized mass increased with PN diameter until it reached about 0.1 parsec, after which it remained constant in the range 0.15 to 0.5 $M_\odot$, as the nebula became optically thin. They concluded that PN shells are ejected at a low velocity during AGB evolution and are continually accelerated throughout the lifetime of the nebula. Combinations of UV and optical region data for six PNs in each cloud suggest PNN masses ranging from 0.58 to 0.66 $M_\odot$ with one object, P25 (which shows strong Fe lines), having a mass of 0.71 $M_\odot$ (Aller *et al.* 1987). The masses of the ionized shells seem to range from approximately 0.03 to roughly 0.5 $M_\odot$, but no data are available on the neutral material. The chemical compositions of these PNs seem to be affected by nuclear processes in the precursor stars. The oxygen depletion in Magellanic Cloud planetaries is even greater than it is in galactic nebulae. Barlow and his associates have obtained extensive spectroscopic data for these engaging objects, whose distances are at least well established, even though their structural properties are difficult to determine (Barlow 1989; Monk, Barlow, and Clegg 1989a,b).

Just what is the scale of the smallest features observable in PNs? The VLA helps greatly here, at least for objects of high flux and surface brightness, but for studies of objects such as NGC 7293, the fine structure will probably be best established with the Hubble Space Telescope (HST). Even before the HST is available for such a project, much structural information can be found for extended bright planetaries by ground-based optical and radio telescopes.

An example in point is the determination of the basic plasma diagnostics and ionic concentrations in N^+, O^+, O^{++}, and S^+ over the surfaces of NGC 40 and NGC 6826 from CCD images secured in the appropriate monochromatic images of the various ionic radiations (Jacoby *et al.* 1987b). In principle, one obtains snapshots of the nebula in the different individual emissions. Conventional spectroscopy usually gives information on line fluxes over a small area or strip of a nebular image. This improved technique is capable of running the gamut of change in chemical composition and plasma parameters (electron density and temperature) over the surface of the nebula. The procedure has the disadvantage of being expensive of telescope time and requiring laborious reductions.

Detailed structural studies of a few bright extended nebulae should be supplemented by radial velocity measurements, obtained with a CCD detector in order to secure a sufficient dynamic range. Chu *et al.* (1986) describe measurements of faint blobs, wisps, and filaments in NGC 6826, where emission from the extended halo of outer radius $64''$ is 1000 times fainter than the brightest components of the lines. The CCD is particularly useful in picking up the emission from very faint, but rapidly moving, wisps or blobs of material. The expansion rate at the outer surface of the nebula is obtained from [N II] and [S II] lines, while other lines help to define the overall picture of expansion. Chu and colleagues concluded that attempts to establish radii versus velocity-of-expansion relationships gave a large scatter because of many stratification effects in a given nebula and large intrinsic differences between different kinds of nebulae as well.

Considerable progress will be possible when detailed spatial plasma diagnostics and surface brightness data are combined with two-dimensional, line-of-sight velocity patterns obtained by the techniques of Reay *et al.* (1983). Such a program carried out for a few carefully selected nebulae for which radio-frequency maps, IR, and UV measurements are also supplied, will permit point-by-point comparisons with realistic theoretical (perhaps time-dependent) models. As for candidates, in addition to NGC 40 and NGC 6826, we will consider NGC 2022, 3242, 6543, 6572, 7662, IC 2165, 4997, and eventually such complex structures as NGC 2440, 6302, and 7027.

Engaging opportunities for spectroscopic studies of bright, high-surface-brightness PNs are offered by modern echelle spectrographs such as the Hamilton spectrograph at the Lick Observatory. With an echelle the spectrum consists of a sequence of parallel clipped sections forming a pattern which is suitable for a CCD detector. With the Hamilton spectrograph a resolution of 0.15 Å is attainable at high efficiency. With the present 800 × 800 CCD chip, six exposures are needed to cover the 100 orders between 3600 and 10,000 Å. Figure 1 shows a raw scan of a portion of one order from an echelle spectrogram of NGC 7027. No corrections for dispersion, field curvature, spectral sensitivity, or division by flat field have yet been made. The intensities of 3954, 3956, and 3961 are of the order of 0.01 to 0.04 on the scale I(Hβ) = 100. Note the separation of the He II 3968 from [Ne III] 3967.

Figure 1. Scan from an echelle spectrogram of NGC 7027.

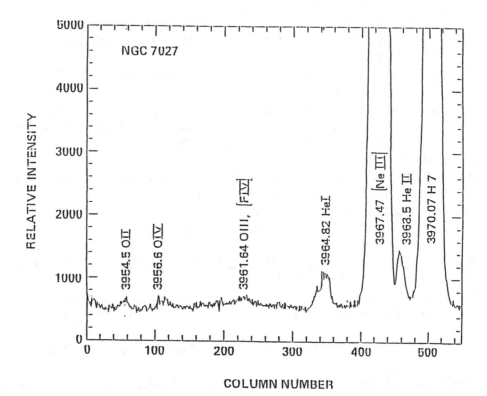

In this quick survey, a number of important topics, including chemical compositions (see e.g., papers in *I.A.U. Symposium No. 103: Planetary Nebulae*, 1983; Kaler 1985; Aller 1984) have been omitted. General accounts of the problems of and physics of gaseous nebulae are given also in Seaton (1960), Aller and Liller (1968), Pottasch (1984), and Osterbrock (1974). Also see papers in *I.A.U. Symposium No. 131: Planetary Nebulae* (1989) for recent developments in this field. It is to be hoped that more astronomers and physicists can be attracted to work on engaging problems in PNs and other gaseous nebulae, and in particular that Bill Liller will return to the fold.

10   ACKNOWLEDGEMENTS

I would like to thank Drs. Balick, Barlow, Iben, Jacoby, Kaler, Köppen, Kwok, Méndez, Peimbert, Pottasch, and Osterbrock, all of whom have sent preprints in advance of publication. I am particularly grateful to Dr. Yervant Terzian, who presented this paper at the Liller Symposium, which I was prevented from attending. Mr. Robert L. O'Daniel, with some technical assistance from Dr. C.D. Keyes, bridged the gap between incompatible word processing systems in preparing the manuscript.

REFERENCES

Abbot, D.C. 1982, *Astrophys. J.*, **259**, 282.

Aller, L.H. 1948, *Astrophys. J.*, **108**, 462.

————. 1984, *Physics of Thermal Gaseous Nebulae*, (Dordrecht: Reidel).

Aller, L.H., Keyes, C.D., and Feibelman, W.A. 1986, *Proc. Natl. Acad. Sci. USA*, **83**, 2777.

Aller, L.H., Keyes, C.D., Maran, S.P., Gull, T.R., Michalitsianos, A.G., and Stecher, T.P. 1987, *Astrophys. J.*, **320**, 159.

Aller, L.H. and Liller, W. 1968, in *Stars and Stellar Systems, Vol. 7: Nebulae and Interstellar Matter*, ed. B.M. Middlehurst and L.H. Aller, (Chicago: University of Chicago Press), p. 483.

Balick, B. 1987, *Sky and Telescope*, **73**, 125.

Balick, B., Bignell, C.R., Hjellming, R.M., and Owen, R. 1987, *Astron. J.*, **94**, 948.

Balick, B. and Preston, H. 1987, *Astron. J.*, **94**, 958.

Barker, T. 1986, *Astrophys. J.*, **308**, 314.

Barlow, M.J. 1989, in *I.A.U. Symp. No. 131: Planetary Nebulae*, ed. S. Torres-Peimbert, (Dordrecht: Kluwer), p. 319.

Barlow, M.J., Morgan, B.L., Standley, C., and Vine, H. 1986, *Mon. Not. Roy. Astron. Soc.*, **223**, 151.

Basart, J.P. and Daub, C.T. 1984, *Bull. Amer. Astron. Soc.*, **16**, 993.

Becklin, E., Neugebauer, G., and Wynn-Williams, C.G. 1973, *Astrophys. Lett.*, **15**, 8.

Beckwith, S., Neugebauer, G., Becklin, E., and Matthews, K. 1980, *Astron. J.*, **85**, 886.

Bentley, A.F., Hackwell, J.A., Grasdalen, G.L., and Gehrz, R.D. 1984, *Astrophys. J.*, **278**, 665.

Bombeck, G., Köppen, J., and Bastian, U. 1986, in *New Insights in Astrophysics, Proc. NASA/ESA/SERC Symposium*, **ESA SP-263**, (Noordwijk: ESA Publ., ESTEC), p. 287.

Castor, J.I., Abbot, D.C., and Kline, R.I. 1975, *Astrophys. J.*, **195**, 157.

Cerruti-Sola, H. and Perinotto, M. 1985, *Astrophys. J.*, **291**, 237.

Cohen, M., Harrington, J.P., and Hess, R. 1984, *Astrophys. J.*, **283**, 687.

Chu, Y.H. 1989, in *I.A.U. Symp. No. 131: Planetary Nebulae*, ed. S. Torres-Peimbert, (Dordrecht: Kluwer), p. 105.

Chu, Y.H., Kwitter, K., Kaler, J.B., and Jacoby, G. 1986, preprint.

Dahari, O. and Osterbrock, D.E. 1984, *Astrophys. J.*, **277**, 648.

Daub, C.T. 1982, *Astrophys. J.*, **260**, 612.

Dopita, M.A., Meatheringham, J., Wood, P.R., Webster, B.L., Morgan, D.H., and Ford, H.C. 1987, *Astrophys. J. Lett.*, **315**, L107.

Drilling, J.S. 1985, *Bull. Amer. Astron. Soc.*, **17**, 594.

Feibelman, W.A. and Kaler, J.B. 1983, *Astrophys. J.*, **269**, 592.

Feibelman, W.A., Hobbs, R.W., McCracken, C.W., and Brown, L.W. 1979, *Astrophys. J.*, **231**, 111.

Gathier, R., Pottasch, S.R., Goss, W.M., and van Gorkom, J.H. 1983, *Astron. Astrophys.*, **128**, 325.

Gathier, R., Pottasch, S.R., and Goss, W.M. 1986, *Astron. Astrophys.*, **157**, 191.

Gathier, R., Pottasch, S.R., and Pei, J.W. 1986, *Astron. Astrophys.*, **157**, 171.

Glassgold, A.E. and Huggins, P.J. 1983, *Mon. Not. Roy. Astron. Soc.*, **203**, 517.

Grauer, A. and Bond, H. 1983, *Astrophys. J.*, **271**, 259.

Grauer, A. and Bond, H. 1984, *Astrophys. J.*, **277**, 211.

Green, R.F., Liebert, J., and Wesemael, F. 1984, *Astrophys. J.*, **280**, 177.

Grewing, M., Bianchi, L., and Cerrato, S. 1986, in *New Insights in Astrophysics, Proc. NASA/ESA/SERC Symposium*, **ESA SP-263**, (Noordwijk: ESA Publ., ESTEC), p. 373.

Hamann, B., Kudritzki, R.-P., Méndez, R.H., and Pottasch, S.R. 1984, *Astron. Astrophys.*, **139**, 459.

Harrington, J.P. and Marionni, P.A. 1976, *Astrophys. J.*, **206**, 458.

Heap, S.R. 1986, in *New Insights in Astrophysics, Proc. NASA/ESA/SERC Symposium*, **ESA SP-263**, (Noordwijk: ESA Publ., ESTEC), p. 291.

Heap, S.R. and Augensen, H.J. 1987, *Astrophys. J.*, **313**, 268.

Hippelein, H.H., Baessgen, M., and Grewing, M. 1985, *Astron. Astrophys.*, **152**, 213.

Iben, I. 1984, *Astrophys. J.*, **277**, 333.

Iben, I., Kaler, J.B., Truran, J.W., and Renzini, A. 1983, *Astrophys. J.*, **264**, 605.

Iben, I. and Tutukov, A.V. 1985, *Astrophys. J. Suppl.*, **294**, 706.

Isaacman, R. 1984, *Astron. Astrophys.*, **130**, 151.

Jacoby, G., Ford, H., Booth, J., and Ciardullo, R. 1987a, *Bull. Amer. Astron. Soc.*, **19**, 712.

Jacoby, G.H., Quigley, P.J., and Africano, J.L. 1987b, *Publ. Astron. Soc. Pacif.*, **99**, 672.

Kaler, J.B. 1983, *Astrophys. J.*, **271**, 188.

————. 1985, *Ann. Rev. Astron. Astrophys.*, **23**, 89.

Kaler, J.B. and Feibelman, W.A. 1985, *Astrophys. J.*, **297**, 724.

Kaler, J.B., Mo, J.-E., and Pottasch, S.R. 1985, *Astrophys. J.*, **288**, 305.

Kwan, J. 1977, *Astrophys. J.*, **216**, 713.

Kwok, S. 1985, *Astrophys. J.*, **290**, 568.

Kwok, S., Purton, C.P., and Fitzgerald, P.M. 1978, *Astrophys. J. Lett.*, **219**, L125.

Langer, G.E., Kraft, R.P., and Anderson, K.S. 1974, *Astrophys. J.*, **189**, 509.

Liller, W. 1955, *Astrophys. J.*, **122**, 240.

Liller, W. and Aller, L.H. 1954, *Astrophys. J.*, **120**, 48.

————. 1963, *Proc. Natl. Acad. Sci. USA*, **49**, 675.

————. 1966, *Mon. Not. Roy. Astron. Soc.*, **132**, 337.

Liller, M.H. and Liller, W. 1968, in *I.A.U. Symp. No. 34: Planetary Nebulae*, ed. C.R. O'Dell, and D.E. Osterbrock, (Dordrecht: Reidel), p. 38.

Liller, W. and Shao, C.-Y. 1970, *Bull. Amer. Astron. Soc.*, **2**, 205.

Livio, M. 1982, *Astron. Astrophys.*, **105**, 37.

Livio, M., Salzman, J., and Shaviv, G. 1979, *Mon. Not. Roy. Astron. Soc.*, **188**, 1.

MacRae, D. and Stock, J. 1954, *Nature*, **173**, 589.

Méndez, R.H., Kudritzki, R.P., Herrero, A., Husfeld, D., and Groth, H.G. 1988, *Astron. Astrophys.*, **190**, 113.

Minkowski, R. and Aller, L.H. 1954, *Astrophys. J.*, **120**, 261.

Monk, D.J., Barlow, M.J., and Clegg, R.E.S. 1989a, in *I.A.U. Symp. No. 131: Planetary Nebulae*, ed. S. Torres-Peimbert, (Dordrecht: Kluwer), p. 354.

————. 1989b, in *I.A.U. Symp. No. 131: Planetary Nebulae*, ed. S. Torres-Peimbert, (Dordrecht: Kluwer), p. 355.

Morris, M. 1981, *Astrophys. J.*, **249**, 572.

Mufson, S.L., Lyon, J., and Marionni, P.A. 1975, *Astrophys. J. Lett.*, **201**, L85.

O'Dell, C.R. and Ball, M.E. 1983, *Astrophys. J.*, **289**, 526.

Okorokov, V.A., Shostov, B.M., Tutukov, A.V., and Yorke, H.W. 1985, *Astron. Astrophys.*, **142**, 491.

Osterbrock, D.E. 1974, *Astrophysics of Gaseous Nebulae*, (San Francisco: Freeman).

Paczyński, B. 1971, *Acta Astron.*, **21**, 417.

Peimbert, M. and Torres-Peimbert, S. 1983, in *I.A.U. Symp. No. 103: Planetary Nebulae*, ed. D.R. Flower, (Dordrecht: Reidel), p. 233.

Pottasch, S.R. 1984, in *I.A.U. Symp. No. 103: Planetary Nebulae*, ed. D.R. Flower, (Dordrecht: Reidel), p. 391.

Reay, N.K., Atherton, P.D., and Taylor, K. 1983, *Mon. Not. Roy. Astron. Soc.*, **203**, 1087.

Reay, N.K., Pottasch, S.R., Atherton, R.D., and Taylor, K. 1984, *Astron. Astrophys.*, **137**, 113.

Renzini, A. 1981, in *Physical Processes in Red Giants*, ed. I. Iben and A. Renzini, (Dordrecht: Reidel), p. 431.

Rodriguez, L.F. and Garcia-Barreto, J.A. 1984, *Rev. Mex. Astron. Astrophys.*, **9**, 153.

Sabbadin, F., Gratton, R.G., Bianchini, A., and Ortolani, S. 1984, *Astron. Astrophys.*, **136**, 181.

Sabbadin, F. 1984, *Mon. Not. Roy. Astron. Soc.*, **210**, 341.

————. 1986, *Astron. Astrophys.*, **160**, 31.

Schaefer, B.E. 1985, *Astrophys. J.*, **297**, 245.

————. 1986, *Astrophys. J.*, **307**, 664.

Schmidt-Voigt, M. and Köppen, J. 1987, *Astron. Astrophys.*, **174**, 211.

Schönberner, D. 1981, *Astron. Astrophys.*, **103**, 119.

————. 1983, *Astrophys. J.*, **273**, 708.

Seaton, M.J. 1960, *Rept. Prog. Phys.*, **23**, 313.

Shaw, R.A. and Kaler, J.B. 1985, *Astrophys. J.*, **295**, 537.

Shields, G.A., Aller, L.H., Keyes, C.D., and Czyzak, S.J. 1981, *Astrophys. J.*, **248**, 569.

Shklovsky, I.S. 1956, *Astron. J. (USSR)*, **33**, 222, 315.

Solf, J. and Weinberger, R. 1984, *Astron. Astrophys.*, **130**, 269.

Terzian, Y. 1987, *Sky and Telescope*, **73**, 128.

Tylenda, R. 1986, *Astron. Astrophys.*, **156**, 717.

Vilkoviskii, H.Y., Kondrateva, L.N., and Tamboseva, I.V. 1983, *Soviet Astron.*, **27**, 194.

Volk, K. and Kwok, S. 1985, *Astron. Astrophys.*, **153**, 79.

Walton, N.A., Reay, N.K., Pottasch, S.R., and Atherton, P.D. 1986, in *New Insights in Astrophysics, Proc. NASA/ESA/SERC Symposium*, **ESA SP-263**, (Noordwijk: ESA Publ., ESTEC), p. 497.

Wood, P.R. and Faulkner, D.J. 1986, *Astrophys. J.*, **307**, 659.

Wood, P.R., Bessell, M.S., and Dopita, M.A. 1986, *Astrophys. J.*, **311**, 632.

Zuckerman, B. 1987, *Sky and Telescope*, **73**, 129.

Zuckerman, B. and Aller, L.H. 1986, *Astrophys. J.*, **301**, 772.

Zuckerman, B., Gilra, D.P., Turner, E.E., Morris, M., and Palmer, P. 1976, *Astrophys. J. Lett.*, **205**, L15.

(*Photo by* Marjorie Nichols)

Yervant Terzian giving a speech in honor of Bill Liller.

STUDIES OF PLANETARY NEBULAE AT RADIO WAVELENGTHS

Yervant Terzian
Cornell University, NAIC, Ithaca, New York, 14853, USA

Abstract. The progress of our understanding of planetary
nebulae is described through radio observations of these
objects. The radio continuum emission is discussed, as well
as the detection of radio recombination lines, and the high
angular resolution radio structure of planetary nebulae.
Observations of the molecular species, H_2, CO and OH are
described, as well as recent detections of neutral hydrogen
at $\lambda 21$ cm surrounding several young and compact nebulae.

1. THE FIRST RADIO DETECTIONS

The progress in our understanding of stellar evolution has
shown that during the late evolutionary stages of stars in the mass
range approximately from 1 to 6 $M_\odot$, the stars eject a substantial
envelope around them, which expands outwards into the interstellar
space. The expanding shells, known as planetary nebulae, receive
significant energy from the hot remnant central stars, and they become
ionized. Typical electron temperatures are $\sim 10^4$ K, and electron
densities $\sim 10^3$ cm^{-3} (Aller and Liller 1968). The expansion velocities
are of the order of 20 to 30 km/sec, and the lifetimes of the visible
nebulosities are $\sim 2 \times 10^4$ years. The remnant central stars become
white dwarfs with a high luminosity of $\sim 10^3$ $L_\odot$, and a surface
temperature of $\sim 10^5$ K.

From the early days of radio astronomy it was clear that an ionized
plasma similar to the visible planetary nebulae should exhibit radio
emission due to the interaction of the free electrons with the ions, a
process known as free-free emission or Bremsstrahlung. Indeed such
radio emission from normal galactic HII regions like the Orion nebula,
Messier 16, Messier 17 and others was detected in the 1950's and
detailed radio studies of these regions had begun in the early 1960's
(Westerhout 1958, Terzian 1965).

During the summer of 1961 the author, as a beginning graduate student,
inspired and encouraged by William Liller and Morton Roberts, attempted
to detect radio emission from planetary nebulae using the then 60-ft
Agassiz radio telescope at Harvard at $\lambda 20$ cm. The results were
inconclusive, but Roger Lynds (1961) using the National Radio Astronomy
Observatory 85-ft Tatel radio telescope succeeded in detecting weak
radio emission from a few planetary nebulae. In 1964 Menon and Terzian

(1965) used the then newly completed 300-ft transit radio telescope at
NRAO and readily measured the flux densities of ten planetary nebulae at
1410 and 750 MHz. These improved measurements opened up the possibility
of checking the recombination theory of emission by comparing the radio
spectrum with the optical Balmer line emission.

This first epoch was concentrated in detecting radio emission from as
many planetary nebulae as possible, and to accurately establish the
radio spectra of these objects. Surveys were performed by Slee and
Orchiston (1965) from Australia for southern planetary nebulae, Terzian
(1966) surveyed the low frequency part of the spectra of planetary
nebulae with the Arecibo 1000-ft radio telescope, Davies et.al. (1967)
made observations at Jodrell Bank of 94 nebulae, and later Rubin (1970),
Higgs (1971a), and Aller and Milne (1972) surveyed a large number of
planetary nebulae, among a few other surveys. In 1971 Higgs (1971b)
compiled a comprehensive catalog of planetary nebulae observed at radio
wavelengths, which included about 100 nebulae with detected radio
emission. Typically planetary nebulae were found to be weak thermal
radio sources, with only a few nebulae having flux densities above 1 Jy.
At centimeter wavelengths most of the detected nebulae have fluxes
between 0.01 and 1 Jy. Figure 1 shows a typical scan of the Ring Nebula
NGC 6720 made at the Arecibo Observatory in 1965 at 430 MHz with a flux
density of 0.5 Jy (Terzian 1968). The strongest nebula with a flux

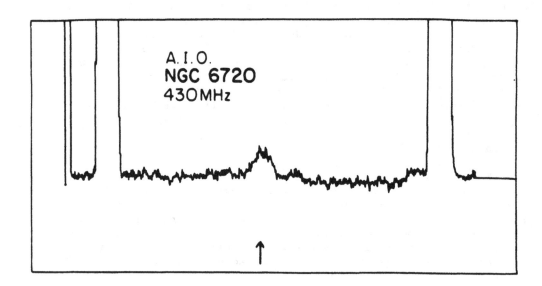

Figure 1. One of the first radio scans of the planetary
nebula NGC 6720 (The Ring Nebula), made with the Arecibo
Observatory 1000-ft radio telescope at 430 MHz, in 1965.

density of ~ 6 Jy at centimeter wavelengths was found to be NGC 7027,
whose spectrum is shown in Figure 2 as compiled by Terzian in 1977 with
additional infrared measurements as reported by Gezari et.al. (1984).

The observations of these nebulae at many radio frequencies clearly
indicated the predicted behavior from ionized hot gases, and showed that

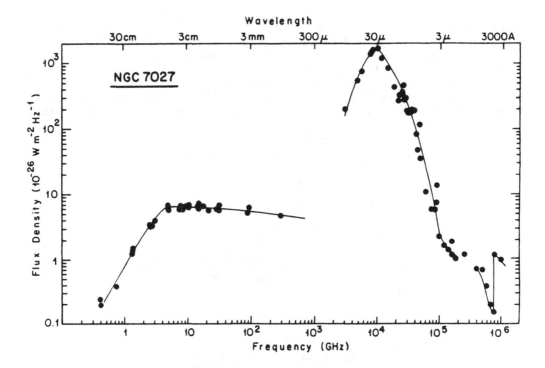

Figure 2. The radio, infrared and optical spectral regions of NGC 7027. The radio spectrum indicates a thermal origin of the emission with the object being optically thin at frequencies higher than ~ 5 GHz, and optically thick at lower frequencies. The infrared spectrum indicates the presence of a warm dust envelope.

these nebulae became optically thick at relatively high frequencies, indicating high electron densities for these objects.

These initial observations also made possible the comparison of the radio emission and the optical Balmer-line emission from these nebulae (Terzian 1966, Pipher and Terzian 1969), and assuming the correctness of the recombination theory it was shown that one could derive interstellar extinction curves at optical wavelengths from the radio and optical measurements.

2. RADIO RECOMBINATION LINES

Radio recombination lines were detected from galactic normal HII regions in the 1960's, and these observations gave important new information on the physical parameters of these objects. It was therefore natural to try and detect these emission lines from planetary nebulae. However, these lines were expected to be very weak and therefore difficult to detect, in part because the lines were supposed to be affected by electron collisional broadening due to the high electron density of these objects. The first attempts to detect such lines were unsuccessful or very marginal, but in 1972 Terzian and Balick

succeeded in detecting the H85α lines from NGC 7027 and IC 418 with good
signal to noise ratios. More recently Churchwell et.al. (1976), and
Walmsley et.al. (1981) have used the Effelsberg 100-meter radio
telescope to detect the H109α and the H76α lines from a number of
planetary nebulae. Figure 3 shows examples of the H76α recombination
line spectra from NGC 7027, IC 418 and NGC 6572 observed by Walmsley
et.al. (1981) as reviewed earlier by one of the authors (Terzian 1980).

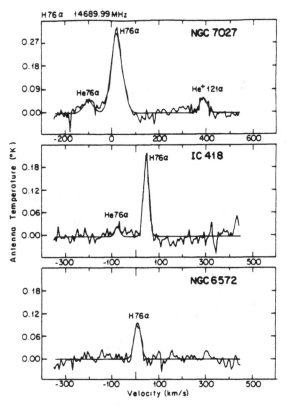

Figure 3. The H76α spectra from NGC 7027, IC 418 and
NGC 6572 observed with the Effelsberg 100-meter radio
telescope.

These observations revealed for the first time the effects of electron
collisional broadening of the recombination lines and made possible the
accurate independent determination of electron densities. For the two
very compact nebulae, NGC 7027 and M1-78, the radio recombination line
observations resulted in   the following electron temperatures and
densities:

NGC 7027: T_e = 14000 K N_e = 3 x 10^4 cm^{-3}

M1-78: T_e = 17500 K N_e = 7 x 10^4 cm^{-3} .

This work has shown that accurate physical parameters can be derived
from Hnα lines provided n is large enough (n > 100) for the lines to be
affected by electron collisional broadening.

3. <u>FINE STRUCTURE IN PLANETARY NEBULAE</u>

        The technical progress of aperture synthesis at radio wave-
lengths made it possible to map the radio brightness distributions of
planetary nebulae with resolutions almost equivalent to optical
resolutions. In 1973 Scott and Balick et.al. independently presented
synthesis radio maps of NGC 7027 with a resolution as fine as 2 arc
seconds. These results revealed a bipolar structure of the ionized gas,
and clearly indicated that a large section of the optical nebula was
severely obscured by dust. Terzian et.al. (1974) published radio
synthesis radio maps of many nebulae at 2.7 and 8.1 GHz where interstellar extinction is completely negligible. These observations were
made with the Green Bank 3-element interferometer with variable
baselines. The results indicated that (a) many nebulae have a bipolar
structure; (b) elliptical shells are present for most of the objects;
(c) central intensity depressions appear in the nebulae for which the
synthesized beam is smaller than the source dimensions; and (d) at the
higher resolutions significant fine structure is evident in the ionized
gas. Figure 4 shows the 8.1 GHz maps of NGC 7027 and NGC 6543 observed
with resolutions of 2 to 3 arc seconds.

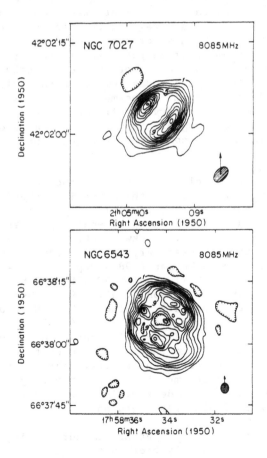

<u>Figure 4.</u> Radio synthesis maps of the planetary nebulae
NGC 7027 and NGC 6543 made with the NRAO 3-element interferometer at 8.1 GHz, in 1973.

The completion of the Very Large Array provided superior aperture
synthesis possibilities in observing the fine structure of planetary
nebulae. Kwok (1985) has observed the shell structure of compact
planetary nebulae with the VLA with resolutions of ~ 0.4 arc second, and
Masson (1986) has measured the angular expansion of NGC 7027 with VLA
observations separated by only 2.8 years. Such results have enabled
Masson to derive a distance to NGC 7027 of 940 ± 200 pc. Terzian (1987)
with the collaboration of C. Bignell, J. Hibbard, and J. van Gorkom have
produced VLA radio maps of several planetary nebulae, including
BD + 30°3639, NGC 7027, NGC 3242, NGC 2392, NGC 6572, and NGC 6210,
during 1983-84. These observations are planned to be repeated during
1988-89 providing a baseline period of five years. It is anticipated
that a comparison of the two images, for each nebula, taken five years
apart will reveal the angular expansion rates, which can be used
together with measured radial velocities of the gas from optical
observations to derive distances to the planetary nebulae.

4. MOLECULES AND NEUTRAL HYDROGEN

One of the most interesting areas of study of planetary
nebulae has been the evolution of red giant stars to proto-planetary
objects and then to the classical stage of the nebulae with ionized,
expanding, visible envelopes. Most of the information on these early
stages of evolution has come from infrared and molecular observations.
The infrared emission has shown that proto-planetary nebulae like
CRL 618, CRL 2688, IRC 10216, and young compact nebulae like NGC 7027,
are associated with dust which emits thermal radiation with a continuum
peak from ~ 30 to 40 μm. The observed infrared spectra for these
objects suggest an equilibrium grain temperature range from ~ 30 K to
~ 200 K, and the grains may be composed of graphite or silicate
particles.

The red giant progenitors of planetary nebulae show a mass loss rate
of the order of $10^{-5}M_{\odot}$ per year, and this rate is probably maintained
for several thousand years. During the initial phases of mass loss the
outer cold material is also composed of molecular species, of which H_2,
CO, CN and OH have been detected. Beckwith et.al. (1978) detected
significant H_2 around compact planetary nebulae and proto-planetary
nebulae, and more recently Zuckerman (1987) reported extended H_2
emission surrounding the Dumbbell nebula NGC 6853 and NGC 2346. CO
emission was first detected by Mufson et.al. (1975), again from young
compact nebulae and proto-planetary nebulae such as NGC 7027, IC 418,
CRL 618, and IRC 10216. Huggins and Healy (1986a,b) have mapped the
spatial distribution of CO around the Helix nebula NGC 7293, NGC 2346
and NGC 6720. These authors find that a signifcant fraction of each
nebula has not yet been ionized by the central star. Observations of
molecular species are of great importance in estimating the fraction of
the gas which is ionized, and in determining more accurately the masses
of the progenitor stars of planetary nebulae.

Attempts to detect neutral hydrogen at λ21 cm surrounding the ionized
gas of planetary nebulae have only been successful very recently.
Rodriguez and Moran (1982) detected an HI absorption feature in the
direction of the young planetary nebula NGC 6302. Rodriguez et.al.

(1985) showed that the HI arises from the dark lane perpendicular to the bipolar ionized nebular features. Altschuler et.al. (1986) have reported HI absorption toward the young nebula IC 4997, and Taylor and Pottasch (1987) have detected HI around IC 418.

The very successful Infrared Astronomical Satellite results have identified numerous galactic late type red giant stars. Many of these are associated with OH sources (Eder et.al. 1987), and these mass ejecting OH/IR stars may be the precursors of planetary nebulae. A recent survey by Payne et.al. of OH from planetary nebulae detected OH from NGC 6302, and Vy 2-2 at 1612 MHz indicating maser action. Vy 2-2 was initially detected by Seaquist and Davis (1983), and these are the only two nebulae with visible ionized envelopes to show dectable OH emission. Figure 5 shows the 1612 MHz OH profiles for these two nebulae.

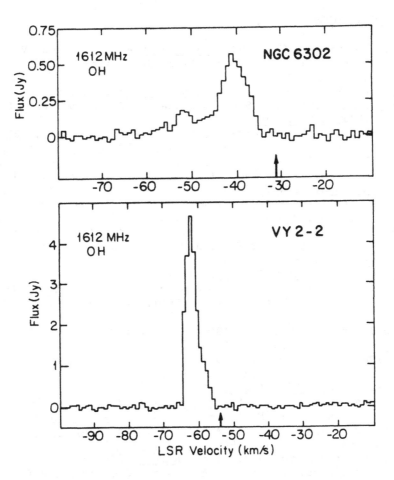

Figure 5. The OH 1612 MHz maser line emission from the planetary nebulae NGC 6302 and Vy 2-2 observed with the NRAO 140-ft radio telescope. The arrows represent the nebular radial velocities obtained from optical observations. The OH lines are blue-shifted indicating that only the near part of the envelope is detected.

It has become apparent that the outer neutral and molecular envelopes of planetary nebulae play an important role in our attempts to understand the late stages of the evolution of stars, and in assessing the influence of the ejected stellar material on the interstellar medium. It has also become clear that information from the visible, infrared, and radio parts of the spectrum are all very essential in order to understand the complex processes of the formation and evolution of planetary nebulae.

This work was supported in part by the National Astronomy and Ionosphere Center, which is operated by Cornell University under a management agreement with the National Science Foundation. The author is grateful to William Liller for introducing him to the pleasures of planetary nebulae.

REFERENCES

Aller, L.H. and Liller, W., 1968, Stars and Stellar Systems, Vol. 7, ed. B. Middlehurst and L.H. Aller, University of Chicago Press, Chicago, 483.

Aller, L.H. and Milne, D.K., 1972, Austr. J. Phys., 25, 91.

Altschuler, D.R., Schneider, S.E., Giovanardi, C. and Silverglate, P., 1986, Ap. J. Lett., 305, L85.

Balick, B., Bignell, C. and Terzian, Y., 1973, Ap. J. Lett., 182, L117.

Beckwith, S., Persson, S.E. and Gatley, I., 1978, Ap. J. Lett., 219, L33.

Churchwell, E., Terzian, Y. and Walmsley, M., 1976, Astron. and Astrophys., 48, 331.

Davies, J.G., Ferriday, R.J., Haslam, C.G.T., Moran, M. and Thomasson, P., 1967, M.N.R.A.S., 135, 139.

Eder, J.A., Lewis, B.M. and Terzian, Y., 1987, Ap. J., (in press).

Gezari, D.Y., Schmitz, M. and Mead, J.M., 1984, NASA Reference Publication 1118, A325.

Higgs, L.A., 1971, M.N.R.A.S., 153, 315.

_____, 1971, "Catalog of Radio Observations of Planetary Nebulae and Related Optical Data", Publ. Astrophys. Branch, N.R.C., Canada, 1, 1.

Huggins, P.J. and Healy, A.P., 1986, Ap. J. Lett., 305, L29.

_____, 1986, M.N.R.A.S., 220, 33P.

Kwok, S., 1985, Astron. J., 90, 49.

Lynds, C.R., 1961, Publ. N.R.A.O., 1, 85.

Masson, C.R., 1986, Ap. J. Lett., 302, L27.

Menon, T.K. and Terzian, Y. 1965, Ap. J., 141, 745.

Mufson, S.L., Lyon, J. and Marionni, P.A., 1975, Ap. J. Lett., 201, L85.

Payne, H.E., Phillips, J.A. and Terzian, Y., 1987, Ap. J., (in press).

Pipher, J.L. and Terzian, Y., 1969, Ap. J., 155, 475.

Rodriguez, L.F. and Moran, J.M., 1982, Nature, 299, 323.

Rodriguez, L.F., Garcia-Barreto, J.A., Canto, J., Moreno, M.A., Torres-Peimbert, S., Costero, R., Serrano, A., Moran, J.M. and Garay, G., 1985, M.N.R.A.S., 215, 353.

Rubin, R.H., 1970, Astron. and Astrophys., 8, 171.

Scott, P.F., 1973, M.N.R.A.S., 161, 35P.

Seaquist, E.R. and Davis, L.E., 1983, Ap. J., 274, 659.

Slee, O.B. and Orchiston, D.W., 1965. Austr. J. Phys., 18, 187.

Taylor, A.R. and Pottasch, S.R., 1987, Astron. Astrophys., 176, L5.

Terzian, Y., 1965, Ap. J., 142, 135.

_____, 1966, Ap. J., 144, 657.

_____, 1968, Planetary Nebulae, ed. D.E. Osterbrock and C.R.
          O'Dell, (Reidel, Dordrecht, Holland), 87.

Terzian, Y. and Balick, B., 1972, Astrophys. Letters, 10, 41.

Terzian, Y., Balick, B. and Bignell, C., 1974, Ap. J., 188, 257.

Terzian, Y., 1977, Sky and Telescope, December, 459.

_____, 1980, Radio Recombination Lines, ed. P.A. Shaver, (Reidel,
          Dordrecht, Holland), 75.

_____, 1987, Sky and Telescope, February, 128.

Walmsley, C.M., Churchwell, E., and Terzian, Y., 1981, Astron.
          Astrophys. 96, 278.

Westerhout, G., 1958, Bul. Astron. Netherl., 14, 215.

Zuckerman, B., 1987, Sky and Telescope, February, 129.

DISCUSSION

J.A. Garcia-Barreto: OH emission from NGC 6302 is very interesting given
the fact that NGC 6302 is considered a proto-planetary nebula. At the
same time it is puzzling. Would you expect the OH to be destroyed as
the central star gets hotter?

Y. Terzian: Indeed OH has not been detected from evolved nebulae. The
OH is probably destroyed by stellar radiation very efficiently.
However, the well established OH/IR stars, which may be the precursors
of planetary nebulae, show strong OH maser emission. In the cases of
NGC 6302 and Vy 2-2, OH is probably being shielded by the intervening
dust.

W. Traub: Do you think you can use the OH maser lines in planetary
nebulae to do distance measurements, via proper motion studies?

Y. Terzian: For the two nebulae where OH has been detected the lines
are not strong enough to detect them with Very Long Baseline Interferometry with available systms. However, some of the OH/IR stars have
strong enough lines that VLBI can be used to measure proper motions.

L.J. Robinson: Does the paucity of binaries in planetary nebulae say
something about the parent stars or is it an observational selection
effect?

Y. Terzian: Most probably it is an observational selection effect that
we do not detect many planetary nebula stars in binary systems. I think
it is almost embarrassing that less than one or two dozen planetary
nebula stars are known to be members of binary systems, compared to the
approximately 1500 planetary nebulae catalogued in our galaxy. We
should try harder!

(*Photo by* Marjorie Nichols)

Josh Grindlay illustrating a point during his talk.

OPTICAL IDENTIFICATIONS OF COMPACT GALACTIC X-RAY SOURCES: LILLER LORE

Jonathan E. Grindlay
Harvard-Smithsonian Center for Astrophysics
Cambridge, MA 02138 USA

Abstract. The optical identification and study of compact galactic x-ray sources was one of the major areas of front-line research at the newly constituted Harvard-Smithsonian Center for Astrophysics in the early 1970s. Bill Liller was the leader of this effort. A brief early history of this work is given together with the connections to currently on-going studies by the author and collaborators. Two of these, the study of ultra-compact x-ray binaries and of galactic plane survey x-ray sources, are described in some detail.

1 INTRODUCTION

By the end of the first decade of work in the fast-growing field of x-ray astronomy, it was clear that optical identification of both galactic and extragalactic x-ray sources was a rewarding and scientifically productive activity. Bill Liller was one of the pioneers in this field, and assumed the role of primary optical astronomer for the burgeoning x-ray group at American Science and Engineering (AS&E), who had just launched the spectacularly successful first x-ray astronomy satellite, *Uhuru*. In the following pages, some of the early history of Liller's contributions is recalled and some of the ways in which his work guided the author's early ventures into this field are described. This review concludes with some of the current areas of active interest in optical studies of galactic x-ray sources and the origin and evolution of x-ray binaries containing either accreting neutron stars or white dwarfs.

2 EARLY HISTORY

X-ray astronomy began in 1962 with the discovery of the brightest x-ray source in the Galaxy, Sco X-1, as well as the diffuse x-ray background (Giacconi *et al.* 1962). The field began with the discovery of two of the principal lures of the field even today: the largely unanticipated consequences of accretion onto compact objects (a neutron star in Sco X-1), and the possible connections to cosmology with the x-ray background radiation, which may arise from distant quasars or hot gas or both. However, even in the early years it was clear that x-ray astronomy could only diverge from its cosmic ray physics parentage and join the astronomical mainstream by first finding, and then studying, the optical counterparts of the newly discovered cosmic x-ray sources. It was (and still is) necessary for the object to be identified with an optical counterpart (if not a radio source) so that its optical magnitude and spectrum could be measured to place it into the astronomical context. X-ray spectra alone were simply not enough, when obtained at the low resolution and sensitivity available, to decode the fundamental nature of the objects.

Thus it was the early involvement of optical astronomers which pushed x-ray astronomy into the fast track. The discovery of the optical counterpart of Sco X-1 by Sandage *et al.* (1966) opened the door for the x-ray astronomers to the world of x-ray binaries as had the earlier identifications of the Crab nebula (Bowyer *et al.* 1964) and M87 (Byram, Chubb and Friedman 1966) opened the door to supernova remnants and active galaxies, respectively.

The optical, and then radio, identifications of Sco X-1 were only partially successful, however. They did not reveal the fundamental nature of the object as a compact binary system. Rather, the variable blue object with its "old nova" type spectrum only suggested possibilities for what was going on inside to produce the prodigious x-ray emission. It was not until the much more detailed x-ray studies with *Uhuru*, the first scientific satellite devoted to x-ray astronomy (launched in December 1970), that the x-ray properties of some of the previously discovered (with short sounding rocket flights) bright galactic x-ray sources became clear: they were magnetized neutron stars. These had only been "discovered" as radio pulsars a few years earlier in 1967. Whereas the radio pulsars were soon realized to be magnetized neutron stars in isolation, the bright galactic x-ray sources were not so quick to reveal their identity as neutron stars accreting gas from a binary companion star. The *Uhuru* discovery of the pulsations and their periodically phase-shifted binary origin, for the x-ray sources Cen X-3 and then Her X-1, showed that these systems were indeed pulsars of a new kind: rotating neutron stars in which the primary energy release mechanism is from accretion of matter onto the compact object (neutron star) rather than electrodynamic losses from pure rotation as in the case of the radio pulsars. However it was not until the discovery of the first optical counterpart of one of these systems, Her X-1, that the overall picture for accreting x-ray binaries could start to become clear. Unlike the blue and erratically variable optical counterpart that had been found for Sco X-1, Her X-1 was much more ready to yield its secrets (and, as usual, further problems). The optical counterpart for Her X-1, the variable star HZ Her, was first noted by Liller (1972).

3   *EARLY OBSERVATIONS*
3.1   *X-ray pulsars*

With the discovery of HZ Her (Liller 1972), x-ray astronomy gained a devoted friend and confidant. Liller, at the Harvard College Observatory, used his knowledge of variable stars and access to the Harvard plate collection as well as the 61-inch telescope at Agassiz Station to good avail. With his discovery of the long-term "off-states", and the discovery that the previously noted "blue variable" HZ Her changed its spectral type from F to A with the 1.7 day period already found by *Uhuru* for the modulation of the 1.24 sec x-ray pulsations, the basic picture became clear: the x-ray emission was due to the transfer of mass from the Roche-lobe of a $\sim$2 M$_\odot$ sub-giant star onto the neutron star in orbit around it. Later observations of the same system would provide the first determination of the mass of a neutron star: approximately 1.4 M$_\odot$. The change in spectral type of the companion star could be understood as a consequence of the x-ray heating effects of the neutron star x-ray source roasting its companion star in x-rays. Additional optical modulation of the system was understood as a direct consequence of the severe tidal distortion of the companion star, which must be pulled into its teardrop-shaped Roche lobe surface by the gravitational field of the neutron star binary companion.

Liller had been pulled into the emerging field of x-ray astronomy by Harvard graduate student Bill Forman and undergraduate Christine Jones, who had already been attracted to the x-ray center of mass: American Science and Engineering, where the *Uhuru* satellite had been planned and was being operated under the direction of Riccardo Giacconi. It was with Christine Jones that Bill Liller was to make his next major optical studies in x-ray astronomy and to begin his continuing close association with southern hemisphere astronomy and Chile. These studies concerned their photometry of HD 77581, the optical counterpart of the 9-day massive x-ray binary system 4U0900-40 (Jones and Liller 1973a), which showed the effects of ellipsoidal light variations due to the tidal distortion of the companion star. Jones and Liller (1973b) also discovered the double-peaked optical variation in the 3.4 day binary 4U1700-37.

3.2   *X-ray bursters*

After the author's discovery of x-ray bursts from the known and bright x-ray source 4U1820-30 in the globular cluster NGC 6624 (Grindlay *et al.* 1976), Liller began a series of projects involving optical studies of globular clusters apparently containing x-ray sources and searches for new clusters. Certainly the crowning achievement came in 1977 with his discovery of the obscured globular cluster containing the remarkable "rapid burster" x-ray source (Liller 1977). This was the result of a well-planned and executed search for heavily reddened clusters in the galactic plane using the newly-available IV-N photographic emulsions. These near-infrared (IR) emulsions required a cumbersome hypersensitization technique, which Liller helped to develop and was among the first to use (along with Victor Blanco) at the 4-m telescope at the Cerro Tololo Inter-American Observatory (CTIO). The discovery of "Liller 1", as his globular cluster came to be called, was significant in that as the second burster to be optically identified it was also in a globular cluster. At the time there was considerable skepticism voiced by the MIT researchers, who had discovered the rapid burster, as to the significance of the association between bursters and globular clusters (*cf.* Lewin 1977). However, it is now clear that all of the 10 luminous x-ray binaries in globular clusters, except that in M15, are x-ray bursters (Grindlay 1985). This indicates that the conditions in globular clusters are especially favorable for the production of x-ray binaries with mass ratios and mass transfer rates suitable for x-ray bursters. The globular cluster sources represent approximately half of the optically identified bursters; the faint blue optical counterparts for systems in the galactic bulge but not in globular clusters, such as the newly identified 4U1915-05 (Grindlay and Cohn 1987, Grindlay *et al.* 1988), may be intimately related to globular clusters, as discussed in the sections below.

3.3 *General identifications*

Certainly it was the lure of the optical observations of globular clusters as the first known sites for x-ray bursters that brought the author into the field of optical studies of compact x-ray sources. Although the author's observing runs were always separate from Liller's, there were many exchanges of finding charts, ideas and plans. Several collaborative projects grew out of these efforts.

The first of these was the identification of the so-called Norma x-ray burster and transient source (Grindlay and Liller 1978), which was – as a matter of historical record – the other burst source originally detected in the discovery of x-ray bursts (the first being that identified by Grindlay *et al.* 1976 with the previously known x-ray source in the globular cluster NGC 6624). The Norma burster was found by Belian *et al.* (1976) using the Vela satellite data but was not possible originally to identify (lacking a precise enough position) with a catalogued x-ray source.

A second joint project was an attempt to find additional highly obscured globular clusters near the galactic center that might be associated with the several x-ray bursters detected there with the SAS-3 satellite. This project produced a mosaic of deep IV-N images (from the CTIO 4-m telescope) of the galactic center and 6 surrounding 1-degree fields as well as a candidate near-IR identification of the galactic center itself (Grindlay and Liller 1979). This search for obscured globular clusters motivated the author's IR-raster scans of the fields of other x-ray burst source positions (Grindlay and Hertz 1981, Hertz and Grindlay 1984a). This program had started with the SAS-3 position ($\sim$30 arcsec) for 4U1728-34 which had been included also on the IV-N plates of the Liller 1 field obtained by both Liller and the author. Examination of these plates suggested a possible diffuse object near the x-ray source position. The more accurate source position later obtained with the Einstein Observatory motivated the IR-scan (at J, H, and K wavebands) of a $\sim$30$\times$30 arcsec region containing the burster and yielded evidence for the most highly obscured globular cluster yet detected (Grindlay and Hertz 1981).

3.4 *Globular clusters*

Finally, it was the early collaborative work with Liller on the structure of globular clusters containing luminous x-ray sources that triggered the author's subsequent work in this field as well. In an effort to search for diffuse gas that could feed a central black hole thought by some (including the author) to power the persistent and burst x-ray emission, concentric aperture narrow-band Hα photometry was performed. This provided evidence for diffuse Hα emission in the cores of several globular clusters (Grindlay and Liller 1977). Although x-ray bursters were even then suspected (on the basis of the thermonuclear flash models just being developed by Joss [1978] and others) to be neutron stars in low mass binary systems, and massive black holes were shortly thereafter ruled out for the x-ray bursters in globular clusters by their significant offsets from the cluster centers (Grindlay 1981), both the evidence for diffuse gas in globular clusters remains (*cf.* Roberts 1988) and central black holes remain as possible contributors to the central surface brightness "cusps" observed in several globular clusters (e.g. Hertz and Grindlay 1985, Djorgovski and King 1984, Lugger *et al.* 1987).

The central cusps are probably indicators of cluster core collapse and the dynamical effects of a central population of "hard" binaries (i.e. those with small separation and thus high orbital velocity). However, the cusp slopes are sufficiently close to the value expected for the stellar distribution around a massive central black hole that this latter possibility cannot yet be ruled out. These hard binaries may include cataclysmic variables (CVs) which contain a white dwarf accreting material from a nondegenerate companion. CVs may give rise to the diffuse Hα emission, as suggested by Grindlay (1988). The definitive test will require higher angular resolution observations closer to the cluster center: a post core collapse cusp will eventually flatten out in its slope to reveal an innermost isothermal region of constant surface brightness, whereas a central black hole will reveal a continual rise, or an even steeper rise, of surface brightness with decreasing radius right into the cluster center (the tidal cutoff radius for a central black hole being completely unresolvable even with the Hubble Space Telescope). Currently the best limits on the central surface brightness slope are for the cusp clusters NGC 6624 and M15. The central surface brightness profile for NGC 6624, a cluster for which Liller previously contributed to surface brightness profile measurements (cf. Canizares *et al.* 1978), is shown in Figure 1.

Figure 1. Central surface brightness profile of NGC 6624 (from Lugger *et al.* 1987). The inner part of the profile is fit with a seeing-convolved power-law cusp of slope −0.77. The outer part of the profile is fit with a normal King (1966) model.

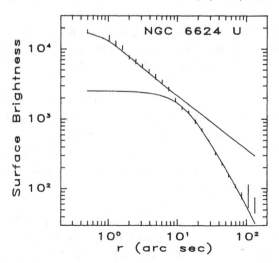

4 RECENT OBSERVATIONS

The globular cluster surface brightness profile shown above is indicative of recent observations of globular clusters and the attempts to study the evolution of globular clusters containing luminous x-ray sources. Other recent observations of x-ray bursters and low luminosity galactic x-ray sources are equally indicative of advances in this field. The following sections briefly describe results obtained for an ultra-compact binary system in the galactic bulge and for the optical identification with newly discovered cataclysmic variables of three low luminosity x-ray sources detected in the Einstein galactic plane survey.

4.1 Identification of 4U1915-05

The x-ray burst source 4U1915-05 is of special interest for several reasons: as the first burster for which an approximate binary period (of 50 min) was found (Walter *et al.* 1982, White and Swank 1982), and as a member of the class of ultra-compact x-ray binaries with doubly-degenerate binary components. In the case of the 4U1915-05 system, the primary is known to be a neutron star by virtue of the x-ray bursts occasionally produced and the general success of the thermonuclear flash model for bursts (*cf.* Lewin and Joss 1981 for a review). The secondary in the system must be a star of $\sim 0.1 M_\odot$ and $\sim 0.1 R_\odot$ in order to fit inside the 50 min binary orbit and fill its Roche lobe radius. In order to supply the mass transfer for the $\sim 10^{37}$ erg/s x-ray luminosity inferred for the persistent emission from the system, the secondary must be degenerate (a He white dwarf) with a hydrogen envelope (Swank, Taam and White 1984). Thus the system is similar to the 11-min ultra-compact binary 4U1820-30 in the globular cluster NGC 6624 (discussed extensively above), for which a binary period was only recently discovered (Stella, Priedhorsky and White 1987). Only with this discovery, did the nature of the 4U1820-30 binary as a doubly-degenerate system, with a neutron star primary and a white dwarf secondary, become clear. Bailyn *et al.* (1988) have shown that it is still not possible to identify the optical counterpart of 4U1820-30 in the core of NGC 6624, because of the extreme crowding of stars in the cluster cusp (the x-ray source is located only ~ 2 arcsec from the center of the radial surface brightness profile shown in Figure 1). In contrast, 4U1915-05 is in the field and thus is accessible to optical identification from the ground. The optical counterpart was suspected (Grindlay 1986) and then confirmed (Grindlay and Cohn 1987, Grindlay *et al.* 1988) in recent CCD photometry from CTIO.

The identification was with a faint (V = 21.0) and blue [(B-V) = +0.4, (U-B) = −0.5] stellar object, which displays dramatic brightness variations on the 50-min orbital period. In fact, observations of the system by a number of observers over a 5-month period reveal that the optical modulation is consistent with a stable period of 50.4567 min and has a light curve shape showing a dip with apparent width and depth both $\sim 30\%$, as shown below in Figure 2.

The optical period is *significantly* different from the best estimates of the x-ray period of 49.9-50.0 min. The x-ray period determinations, reported in the original period-discovery papers of Walter *et al.* (1982) and White and Swank (1982),

Figure 2. Folded light curve for 4U1915-05 (from Grindlay *et al.* 1988).

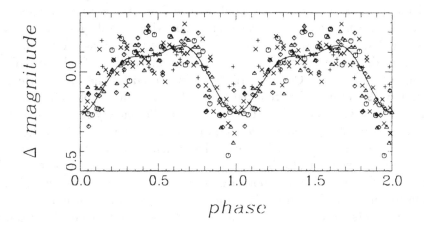

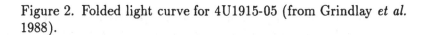

are derived from the times of the irregular x-ray dips (reductions of some 30% in intensity due to both scattering and absorption of x-rays from obviously clumped gas in the binary system). Although the x-ray dips show both phase jitter and some evidence of period wandering, they are always observed with apparent periods in the range 49.9-50.1 min and thus are not consistent with the optical period.

The fact that the optical period is longer than the x-ray period may indicate that the system is modulated by a third body: a triple companion in a retrograde orbit (at approximately 2/3 of the 4 day synodic period suggested by the $\sim$ 1% difference between the x-ray and optical periods) could modulate the mass transfer onto the neutron star and give rise to the x-ray dips at the slightly shorter period (*cf.* Grindlay *et al.* 1988). As discussed in that paper, the hierarchical triple model is also consistent with the previous interpretation of the long-term (199 day) period (Priedhorsky and Terrell 1984) as due to precession (Grindlay 1986). The model is also supported by the general arguments for the formation of field bursters – especially triples – by tidal capture in globular clusters since disrupted (Grindlay 1984, 1988).

The *retrograde* nature of the triple companion, if indeed that is the correct interpretation of the difference between the x-ray dip and optical periods, is especially important since retrograde triples (only) can be formed by tidal capture onto pre-existing compact binaries in globular clusters (Bailyn 1987a) but are most unlikely to form from the evolution of "primordial" triples. This is because primordial triples, which are defined here as those triples which have formed at a common epoch of star formation possibly including fission, are almost certainly prograde systems, since they arise from a common rotating gas cloud.

4.2 *Identification of GPX sources with cataclysmic variables*

As a final example of recent activity in the identification and study of compact galactic x-ray sources, our on-going study of the low-luminosity sources detected serendipitously by the Einstein X-ray Observatory is discussed. The galactic plane survey was conducted by examining the serendipitous sources detected in Einstein observations carried out within 15 degrees of the galactic plane and subject to a number of selection criteria, such as the absence of known bright or diffuse (e.g. supernova remnants) sources in the field of the Einstein Imaging Proportional Counter (IPC) detector used for the survey (Hertz and Grindlay 1984b; hereafter referred to as HG). The major objective of the galactic plane survey was, and still is, to constrain the luminosity functions of ordinary stars (as coronal sources) and low luminosity accretion sources in the Galaxy. The latter objective has in fact been the principal one (coronal source studies have been carried out also by a number of other surveys, such as those of star formation regions) and has motivated our follow-up work. Our main aim has been to constrain the luminosity function and space distribution of low luminosity ($\leq 10^{35}$ erg/sec) soft x-ray (i.e. in the $\sim 0.3 - 3$ keV band of the Einstein Observatory) sources containing compact objects. While most of these sources are expected to be cataclysmic variables (CVs), some may be binaries with neutron stars accreting at low mass transfer rates.

From the initial survey results (Hertz 1983, HG), it appeared that a significant fraction ($\sim 25\%$) of these sources, with x-ray fluxes typically $\sim 0.01 - 0.1$ *Uhuru* Flux Units (UFU), would be CVs. Of the remaining sample, our original predictions were that two-thirds would be coronal sources, with the predominant contributors being M-dwarf stars, and the remaining third would be active galactic nuclei (AGNs) shining through the galactic plane.

In addition, the original sample of galactic plane x-ray (GPX) sources showed a striking asymmetry between the northern and southern skies, with nearly a factor of two more presumed binary accretion sources in the southern sample after removing the statistically expected number of stellar coronal sources and AGNs (HG). Since the center of the Galaxy is located in the southern sky, this observed excess suggests that a Population II component (i.e. one associated with the galactic bulge and halo rather than the disk) may be a significant contributor to the low luminosity x-ray source distribution of the Galaxy. The fact that the distribution of optically detected novae, a type of CV, is strongly peaked towards the center of M31 (Ciardullo *et al.* 1987), further supports the interpretation that the north-south asymmetry in the GPX sample is due to an increased number of CVs towards the galactic center.

In order to test the inferred source populations, it is of course necessary to identify them optically. Accordingly, we have carried out a program of spectroscopic identification of northern GPX source fields using the 1.5-m and MMT telescopes on Mt. Hopkins (Hertz and Grindlay 1988). We have also initiated a program of both CCD photometry and follow-up spectroscopy for the southern GPX sample using the facilities at CTIO. In the work on the northern sample, which was restricted to GPX fields north of declination $-20°$, we have observed fields for which bright stars are not contained in the typical ~ 1 arcmin IPC error circles and thus the source is not an a priori candidate for being a coronal source. The fields were chosen such that no star in the error circle was bright enough to yield a ratio of x-ray to optical flux less than 0.01. Since virtually all stellar coronal emission sources (except M-dwarfs) would be expected to be excluded by this criterion, the optical identification survey was intentionally planned to maximize the chances of finding low luminosity galactic accretion sources rather than single normal stars. Nevertheless, no additional CVs were found (beyond the one system noted in the original northern GPX sample discussed by HG), and the conclusion emerged that the northern GPX sample is dominated by coronal sources and AGNs, which "leak" through the galactic plane. The northern sample has only been observed spectroscopically down to 16th magnitude (virtually complete for all fields), although several fields have been observed with the MMT to fainter magnitudes. Thus, the possibility remains that some of the sources for which M-dwarfs or AGN counterparts were not found are fainter CVs.

This possibility is made all the more likely by the fact that for the southern GPX sample, our first spectroscopic follow-up to two runs of UBV CCD photometry at CTIO has yielded three new CVs (Grindlay et al. 1987). The candidates were identified first as uv-excess objects in color-color diagrams of the stars in the IPC error circle from the CCD photometry. An example of such a diagram, with the candidate marked, is shown below in Figure 3.

Figure 3. Color-color diagram for field of GPX 84, with counterpart marked.

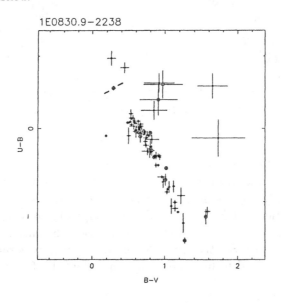

Figure 4, below, shows the spectrum of the candidate counterpart to GPX 84, obtained in May 1987 at the CTIO 4-m. Clearly the object is a CV, probably of the AM Her type given the He II emission in its spectrum. Spectra of the other two new CVs found were of lower excitation, with only the hydrogen Balmer lines seen strongly in emission. From the equivalent widths of the Balmer emission lines in all three objects, approximate distances may be derived using the correlations of Patterson (1984). The resulting distances of at least 2 kpc are greater than those for virtually all optically selected CVs, suggesting that the x-ray selection afforded by the GPX survey can indeed sample much larger volumes for determining the distribution of CVs in the Galaxy. The discovery of three CVs (and the detection of a fourth previously identified CV), out of a sample of 8 southern GPX sources examined spectroscopically, suggests that the southern excess of GPX sources is probably dominated by CVs. Extensive CCD photometry and 4-m spectroscopy of southern GPX fields at CTIO in June 1988 have resulted in another possible CV identification.

Figure 4. Spectrum of the uv-excess object in the field of GPX 84.

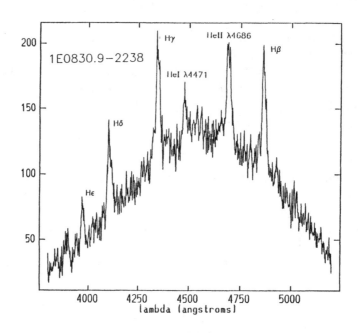

5   CONCLUSIONS

It is clear that Bill Liller started a good thing, with his early efforts to optically identify x-ray sources in the Galaxy. Not only did his work result in important new discoveries, but it also undoubtedly helped to convince Riccardo Giacconi and his collaborators that a move from American Science and Engineering to the newly formed Harvard-Smithsonian Center for Astrophysics made good sense. It also certainly helped to lure the author into the field, for which he is grateful.

This work was supported by Bill Liller and, more recently, by grants NSF AST 84-17846, and NASA NAGW-624 and NAS8-30751.

REFERENCES

Bailyn, C. 1987a, Ph.D. Thesis, Harvard University.

———. 1987b, *Astrophys. J.*, **317**, 737.

Bailyn, C., Grindlay, J., Cohn, H. and Lugger, P. 1988, *Astrophys. J.*, **331**, 303.

Bailyn, C. and Grindlay, J. 1987, *Astrophys. J. Lett.*, **316**, L25.

Belian, R., Conner, J., and Evans, D. 1976, *Astrophys. J. Lett.*, **206**, L135.

Bowyer, S., Byram, E.T., Chubb, T.A., and Friedman, H. 1964, *Nature*, **201**, 1307.

Byram, E.T., Chubb, T.A., and Friedman, H. 1966, *Science*, **152**, 66.

Canizares, C., Grindlay, J., Hiltner, W., Liller, W., and McClintock, J. 1978, *Astrophys. J.*, **224**, 39.

Ciardullo, R., Ford, H.C., Neill, J.D., Jacoby, G.H., and Shafter, A.W. 1987, *Astrophys. J.*, **318**, 520.

Djorgovski, S. and King, I. 1984, *Astrophys. J. Lett.*, **277**, L49.

Giacconi, R., Gursky, H., Paolini, F., and Rossi, B.B. 1962, *Phys. Rev. Lett.*, **9**, 439.

Grindlay, J.E. 1981, in *X-ray Astronomy With the Einstein Satellite*, ed. R. Giacconi, (Dordrecht: Reidel), p.79.

———. 1984, *Adv. Space. Res.*, **3**, No. 10, 19.

———. 1985, in *Proc. Joint US-Japan Seminar on Galactic and Extragalactic Compact X-ray Sources*, eds. Y. Tanaka and W.H.G. Lewin, (Tokyo: ISAS), p. 215.

———. 1986, in *Origin and Evolution of X-ray Binaries, Proc. NATO Conf. at Tegernsee*, eds. W. Brinkman, W. Lewin, and J. Trumper, ASI Vol. 267, p. 25.

———. 1988, in *IAU Symposium 126, Globular Cluster Systems in Galaxies*, eds. J.E. Grindlay and A.G.D. Philip, (Dordrecht: Kluwer), p. 347.

Grindlay, J.E., Bailyn, C.D., Cohn, H., Lugger, P.M., Thorstensen, J.R., and Wegner, G. 1988, *Astrophys. J. Lett.*, **334**, L25.

Grindlay, J.E. and Cohn, H. 1987, *IAU Circular*, **4393**.

Grindlay, J., Cohn, H., Lugger, P., and Hertz, P. 1987, *IAU Circular*, **4408**.

Grindlay, J.E., Gursky, H., Schnopper, H., Parsignault, D.R., Heise, J., Brinkman, A.C., and Schrijver, J. 1976, *Astrophys. J. Lett.*, **205**, L127.

Grindlay, J.E. and Hertz, P. 1981, *Astrophys. J. Lett.*, **248**, 47.

Grindlay, J.E. and Liller, W. 1977, *Astrophys. J. Lett.*, **216**, L105.

———. 1978, *Astrophys. J. Lett.*, **220**, L127.

———. 1979, *Adv. Space Expl.*, **3**, 83.

Hertz, P. 1983, Ph.D. Thesis, Harvard University.

Hertz, P. and Grindlay, J.E. 1984a, *Astrophys. J. Lett.*, **282**, 118.

———. 1984b, *Astrophys. J.*, **278**, 137, (HG).

———. 1985, *Astrophys. J.*, **298**, 95.

———. 1988, *Astron. J.*, **96**, 233.

Jones, C. and Liller, W. 1973a, *Astrophys. J. Lett.*, **184**, L121.

———. 1973b, *Astrophys. J. Lett.*, **184**, L65.

Joss, P.C. 1978, *Astrophys. J. Lett.*, **225**, L123.

King, I.R. 1966, *Astron. J.*, **71**, 64.

Lewin, W.H.G. 1977, *Ann. NY Acad. Sci.*, **302**, 210.

Lewin, W.H.G. and Joss, P.C. 1981, *Space Sci. Rev.*, **28**, 3.

Liller, W. 1972, *IAU Circular*, **2415** and **2427**.

———. 1977, *Astrophys. J. Lett.*, **213**, L21.

Lugger, P.M., Cohn, H., Grindlay, J.E., Bailyn, C., and Hertz, P. 1987, *Astrophys. J.*, **320**, 482.

Patterson, J. 1984, *Astrophys. J. Suppl.*, **54**, 443.

Priedhorsky, W. and Terrell, J. 1984, *Astrophys. J.*, **280**, 661.

Roberts, M. 1988, in *IAU Symposium 126, Globular Cluster Systems in Galaxies*, eds. J.E. Grindlay and A.G.D. Philip, (Dordrecht: Kluwer), p. 411.

Sandage, A.R., Osmer, P., Giacconi, R., Gorenstein, P., Gursky, H., Waters, J., Brandt, H., Garmire, G., Sreekantan, B.V., Oda, M., Osawa, K., and Jugaku, J. 1966, *Astrophys. J.*, **146**, 316.

Stella, L., Priedhorsky, W., and White, N. 1987, *Astrophys. J. Lett.*, **312**, L17.

Swank, J., Taam, R., and White, N. 1984, *Astrophys. J.*, **277**, 274.

Walter, F., Bowyer, S., Mason, K., Clarke, J., Henry, J., Halpern, J., and Grindlay, J. 1982, *Astrophys. J. Lett.*, **253**, L67.

White, N. and Swank, J. 1982, *Astrophys. J. Lett.*, **253**, L61.

AGES OF GLOBULAR CLUSTERS DERIVED FROM BVRI CCD PHOTOMETRY

Gonzalo Alcaino
Instituto Isaac Newton
Ministerio de Educación de Chile
Santiago, Chile

1 INTRODUCTION

Some years ago, when Yervant Terzian and I (and others) were doing graduate work at the university where Phyllis Lugger and Haldan Cohn now profess, I gave careful thought to the type of research that I might carry out when I returned to my home in Chile. At that time the Cerro Tololo Interamerican Observatory (CTIO) and the European Southern Observatory (ESO) were just being started. I was primarily interested in photometry, especially due to the help and encouragement of James Cuffey, who was then on the Indiana University faculty.

The beautiful work on globular clusters that had recently been done by Arp, Baum, and Sandage at the Mount Wilson and Palomar Observatories had impressed me greatly, and so I decided to stop off in Pasadena on my way home and talk with these experts. It was then and there that the decision was made: do photometry on the many globular clusters that lie at negative declinations (65% are south of $-20°$).

While the first telescopes at Tololo were not large, globular cluster photometry in the Southern Hemisphere was a virgin field, and almost any cluster that I worked on was being studied for the first time. It was possible to reach below the horizontal branch, home of the RR Lyrae stars, in most clusters, thereby permitting me to derive reasonably reliable distances and interstellar reddenings for these objects. Also, approximate metallicities could be determined. This work culminated in the publication of the book, *Atlas of Galactic Globular Clusters with Colour Magnitude Diagrams* (Alcaino 1973).

The early photometry was carried out in two steps. First, a dozen or so stars were measured photoelectrically (UBV usually), covering as wide a range of color and magnitude as possible. Then photographs were taken and star images measured with a suitable photometer such as had been developed by Cuffey. On some telescopes a thin-wedge prism could be placed before the objective thereby producing a second, much fainter image of each star. Thus, in a boot-strap manner, the relatively bright photoelectric sequence could be extended to fainter magnitudes.

Then came charge coupled devices (CCDs). Suddenly it became possible to carry out photometry with substantially higher precision owing to the stability and the linearity of these marvelous detectors, especially at low light levels. Also, the sensitivity range extended beyond 1 micron making it possible to work with ease in more colors than B and V, which had been the workhorse wavelength bands for many years.

Our William, whose 60*th* birthday we celebrate with this volume, moved permanently to Chile in mid 1981, at a time when CCDs were still not available to users at the Chilean observatories of CTIO and ESO. We had known each other since 1968 when we both happened to be observing at Tololo. Because of his interest in globular clusters, we began to collaborate, initially using the now old-fashioned techniques, but then changing over to CCDs when they became available.

In this paper I shall describe the fruits of our recent work. Let me emphasize at the outset that William's name should appear as a coauthor, but he insisted most firmly that it be omitted in this volume dedicated to him. However, you should understand that he has been coauthor with me now on over 40 papers; only because I deeply cherish our close friendship do I assent to his wishes.

2   CURRENT STATUS OF RESEARCH

In his summary of the current status of deep two-color (B,V) photometry of galactic globular clusters, Charles Peterson (1986) notes that color-magnitude diagrams (CMDs) reaching down to the main sequence have been obtained for 36 globular clusters in the Galaxy (Table 1). The primary motivation for this research has been the determination of the ages of the clusters found by matching the observed V, (B−V) diagram to the theoretically derived luminosity-temperature relationship converted to the observed parameters. Initially, it was not clear if globular clusters had a substantial spread in age which would imply a gradual formation of the halo of the Galaxy (see, for example, Carney 1980, Demarque 1980), or if the ages were all similar which would lead one to conclude that there was a rather abrupt galactic collapse (see Sandage 1982). However, improved work of recent years, both observational and theoretical, has made it clear that there is a small spread in the ages of globular clusters, perhaps no more than 1 billion years, and Burstein (1985) has advocated an average age somewhere between 14 and 17 Gyr.

One of the most important aspects of this result is that it bears directly on the age of the Universe and the value of the Hubble constant since globular clusters are among the oldest known objects in existence. Therefore, establishing firmly and accurately the age of the oldest clusters would set an upper limit on the Hubble constant, the exact value of which continues to be very much under discussion at the present time.

One of the weaknesses of the extremely popular and often-used BV photometric system is that metallic line absorption in the blue and violet can be considerable.

Table 1
Clusters with Photometry to the Turnoff Magnitude

Cluster	V_{HB}	E(B−V)	ΔV_{TO-HB}[1]	N	[Fe/H][2]
NGC 104	14.05	0.04	3.61 ± 0.08	4	−0.85
NGC 288	15.3	0.03	3.65 ± 0.09	4	−1.31
NGC 362	15.43	0.04	3.55	−	−1.06
NGC 1904	16.15	0.01	3.34	2	−1.45
NGC 2298	16.09	0.08	3.36 ± 0.05	2	−1.66 :
NGC 2808	16.11	0.22	3.58 ± 0.03	2	−1.09
NGC 3201	14.72	0.21	3.33	−	−1.19
Pal 4	20.63	0.02	3.27 ± 0.07	2	−2.15
NGC 4590	15.63	0.03	3.60	−	−2.11
NGC 5053	16.63	0.01	3.35	−	−2.15
NGC 5139	14.50	0.11	3.54 ± 0.04	2	−1.64
NGC 5272	15.66	0.00	3.4	−	−1.43
NGC 5466	16.58	0.00	3.32	−	−1.6
Pal 5	17.40	0.03	3.32 ± 0.02	2	−1.16
NGC 5904	15.15	0.03	3.41	−	−1.31
NGC 6121	13.36	0.39	3.39 ± 0.21	2	−1.23
NGC 6171	15.63	0.35	3.67 ± 0.09	2	−0.94
NGC 6205	14.85	0.03	3.30 ± 0.11	2	−1.41
NGC 6229	18.08	0.10	3.45	−	−1.25
NGC 6254	14.65	0.26	3.75	−	−1.39
NGC 6341	15.05	0.02	3.15	−	−2.05
NGC 6352	15.2	0.23	3.6	−	−0.8
NGC 6362	15.34	0.10	3.41	−	−0.96
NGC 6397	12.95	0.18	3.75 ± 0.05	2	−2.24
NGC 6535	15.8	0.36	3.7	−	−1.4*
NGC 6656	13.97	0.35	3.4	−	−1.55
NGC 6752	13.72	0.04	3.52 ± 0.02	3	−1.23
NGC 6809	14.33	0.08	3.7	−	−1.55
Pal 11	17.3	0.34	3.2	−	−
NGC 6838	14.41	0.27	3.55	−	−0.79
NGC 7006	18.79	0.00	3.56	−	−1.32
NGC 7078	15.83	0.11	3.45 ± 0.12	−	−2.01
NGC 7089	16.05	0.06	3.45 ± 0.20	2	−1.48
NGC 7099	15.05	0.02	3.35	−	−2.32
Pal 12	17.1	0.02	3.2	−	−1.0*
Pal 13	17.70	0.00	3.29	−	−1.8

[1]Mean is 3.457 ± 0.027 (s.d.m.).
[2]Pilachowski (1984) unless indicated with an asterisk, for which [Fe/H] comes from a calibration with $\Delta V = 1.4$.

Thus, the interpretation of observations in the context of stellar evolution theory rests heavily on model stellar atmospheres. These models are needed to predict the effects of metallic line absorption in stars where the metallicity is often not well known. The metallicity of globular clusters, usually expressed by the parameter [Fe/H], has a range of over two orders of magnitude, making a precise knowledge of absorption line effects imperative.

Fortunately, VandenBerg and Bell (1985) have now carried out the difficult calculations needed to predict the location of isochrones in longer wavelength bands where there is much less metallic absorption than in the blue. This work came largely as a response to the appearance of the modern generation of electronic detectors, especially the CCD, which has made it possible to carry out improved photometry at magnitudes fainter than photographic limits and at wavelengths extending into the near infrared. Besides the obvious advantage of reducing uncertainties arising from poorly known metallicities, use of the red and infrared wavelengths (R and I bands) makes possible an enlarged color baseline which enhances effects seen less clearly with a smaller range in color. An additional benefit of this multicolor approach is that observational uncertainties are reduced by having several independently derived CMDs of the same cluster. With several separate evaluations of the age of a single cluster, one not only can assess more reliably the accuracy of the final result, but also can derive ages with a higher precision than attained previously or with only two colors.

Consequently, for the past several years William and I have had underway a program of BVRI photometry of globular clusters using CCDs and new powerful reduction software. The main thrust of this program is to concentrate on relatively nearby clusters so that we can reach well below the main sequence turnoff with medium-sized telescopes.

We have now completed BVRI reductions on five clusters: NGC 104 (47 Tuc), NGC 2298, NGC 5139 (ω Cen), NGC 6121 (M4), and NGC 6362. See Table 2 for a summary of the cluster parameters. All observations were made with the excellent CCD system of the 1.54 meter Danish telescope at ESO-La Silla. An additional eight globular clusters have now been observed in these four colors with a CCD at the 2.2 meter Max Planck telescope at La Silla (see Table 3), and these data are currently being reduced at the computer center at La Silla and analyzed at the Isaac Newton Institute in Santiago.

In order to minimize the errors that arise from comparing cluster fields with standards in other parts of the sky, we have set up photoelectric standards in the same cluster fields observed with the CCD. Thus, the effects of inaccurately known or varying atmospheric extinction are totally avoided. Because we can use smaller telescopes to set up the standards (most often the ESO 1 meter reflector), valuable large telescope time is not wasted moving back and forth between widely separated fields.

Table 2
Globular Clusters Studied with BVRI CCD Photometry

Cluster	b°	E(B−V)	[Fe/H]	$d_\odot$ (kpc)
NGC 104 (47 Tuc)	−44.9	0.04	−0.44	4.6
NGC 2298	−16.0	0.11	−1.41	12.4
NGC 5139 (ω Cen)	15.0	0.11	−1.1, −2.2	5.2
NGC 6121 (M4)	16.0	0.35	−1.30	2.2
NGC 6362	−17.6	0.12	−0.9	7.2

CCD frames obtained with ESO – Danish 1.54 m telescope.
(Data from Harris and Racine 1979).

Table 3
Globular Clusters with BVRI CCD Data now being Analyzed at the Isaac Newton Institute

Cluster	b°	E(B−V)	[Fe/H]	$d_\odot$ (kpc)
NGC 1851	−35.0	0.07	−1.29	10.9
NGC 1904	−29.3	0.01	−1.58	13.3
NGC 2808	−11.3	0.22	−1.09	9.4
NGC 3201	8.6	0.21	−1.26	5.1
NGC 4372	−9.9	0.45	−1.7	5.1
NGC 4590	36.0	0.03	−2.04	9.6
NGC 5946	4.2	0.56	−1.5	10.1
NGC 6139	6.9	0.68	−1.27	9.5

CCD frames obtained with ESO – Max Planck 2.2 m telescope.
(Data from Harris and Racine 1979).

3 OBSERVATIONS

The CCD camera with which the clusters listed in Table 2 were observed, is located at the Cassegrain focus of the Danish 1.54 meter telescope and consists of 512×320 pixels. Each 30×30 micron pixel corresponds to an area on the sky of $0.47'' \times 0.47''$; thus, the total field is $4.0' \times 2.5'$. A number of exposures in each of the BVRI passbands were obtained for each cluster with exposure times ranging from a few seconds to several minutes, in order to cover the full range of magnitudes without image saturation. A brief summary of the observations appears in Table 4. Field positions were chosen, following careful inspection of deep photographs, as the best compromise between maximum cluster membership and

Table 4
Number of Frames Observed in Five Clusters

Cluster	Fields	B	V	R	I	Total
47 Tuc	1	10	10	–	11	31
NGC 2298	1	11	13	12	12	48
ω Cen	2	46	48	41	44	179
M4	1	10	10	–	11	31
NGC 6362	1	23	17	24	14	78

workable contamination caused by crowding of images.

Originally, reductions were carried out at La Silla using a conventional photometric program (HP/IHAP). Recently, however the MIDAS/INVENTORY software employing point spread function techniques has become available, and because of the better results obtainable, we are now using MIDAS exclusively. Precise color equations have now been established for the CCD telescope system based on 140 BVRI frames and a total of 40 standard stars with a large range of color.

4   THE RESULTS

The CMDs for the three clusters reduced with MIDAS (47 Tuc, M4, and ω Cen) are shown in Figures 1-3. As an example, we will discuss in somewhat more detail the results for 47 Tuc.

Two hundred and two stars were observed in a $4.0' \times 2.5'$ field and the MIDAS/INVENTORY reductions yielded the V vs. B−V, V vs. V−I, and V vs. B−I CMDs shown in Figures 1a-c. These diagrams also include the theoretical isochrones of VandenBerg and Bell (1985) for ages ranging from 8 to 18 Gyr. From the figures we can see that the brightest stars investigated in our field are four red horizontal branch stars at a mean magnitude of V = 14.07. While no red giants are present in the field, the subgiant branch and the main sequence are well defined in all diagrams. The magnitudes of the main sequence turnoff points in the three colors all fall very close to $V_{TO} = 17.60 \pm 0.10$. The main sequence color turnoffs lie at B−V = 0.56 ± 0.02, V−I = 0.68 ± 0.02, and B−I = 1.24 ± 0.02. Hesser and Hartwick (1977) and Lee (1977) found that the horizontal branch lies at $V_{HB} = 14.05 \pm 0.05$. Thus the difference between the turnoff magnitude and horizontal branch magnitude is $\Delta V_{TO-HB} = 3.55 \pm 0.20$, in agreement with the mean value derived by Peterson (1986) for the clusters listed in Table 1, namely $\Delta V_{TO-HB} = 3.457 \pm 0.027$.

Our main sequence turnoff values and fiducial points for the V vs. B−V CMD can be compared directly with the results obtained from the SIT Vidicon camera at the 4 meter CTIO telescope by Harris, Hesser, and Atwood (1983a,b). The comparison is excellent. For the turnoff point they deduce $V_{TO} = 17.55 \pm 0.20$ and B−V = 0.58 ± 0.02, and all of the fiducial points agree within ± 0.02 mag.

Figure 1. The observed V vs. B−V, V vs. V−I, and V vs. B−I
color magnitude diagrams of 47 Tuc, fitted to the isochrones of
VandenBerg and Bell (1985) for Y = 0.2, Z = 0.006, ([Fe/H] =
−0.49), α = 1.65, and ages 8-18 Gyr. The isochrones were shifted
to represent a cluster with (m-M)$_V$ = 13.2 and with: (a) E(B−V) =
0.04, (b) E(V−I) = 0.05, and (c) E(B−I) = 0.09.

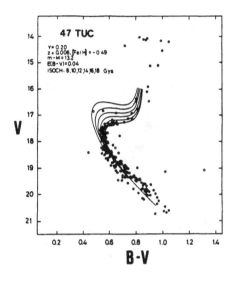

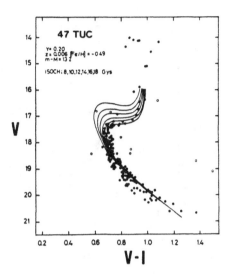

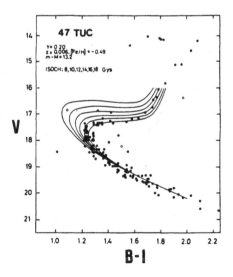

Figure 2. The observed V vs. B−V, V vs. V−I, and V vs. B−I color magnitude diagrams of M4, fitted to the isochrones of VandenBerg and Bell (1985) for Y = 0.2, Z = 0.001, ([Fe/H] = −1.27), α = 1.65, and ages 8-18 Gyr. The isochrones were shifted to represent a cluster with (m-M)$_V$ = 12.7 and with: (a) E(B−V) = 0.41, (b) E(V−I) = 0.38, and (c) E(B−I) = 0.80.

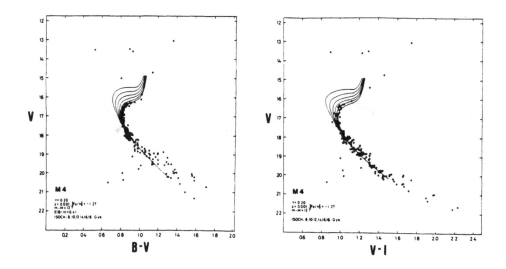

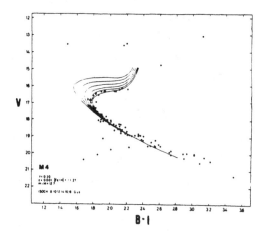

Figure 3. The observed V vs. B−V, V vs. V−I, and V vs. B−I color magnitude diagrams of ω Cen, fitted to the isochrones of VandenBerg and Bell (1985) for Y = 0.2, Z = 0.001, ([Fe/H] = −1.27), α = 1.65, and ages 8-18 Gyr. The isochrones were shifted to represent a cluster with (m-M)$_V$ = 14.0 and with: (a) E(B−V) = 0.15, (b) E(V−I) = 0.16, and (c) E(B−I) = 0.33.

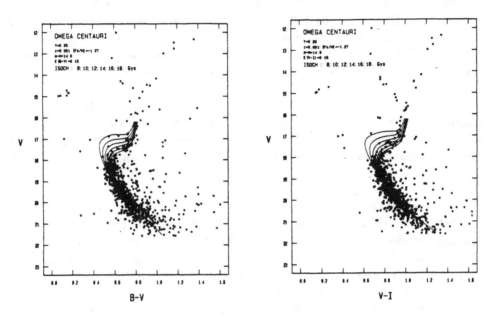

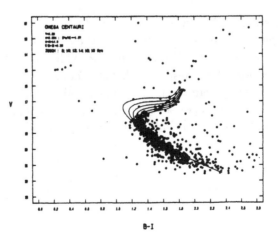

The VandenBerg and Bell (1985) BVRI isochrones are presented with the helium content alternatives of Y = 0.2 and Y = 0.3, and in the metal rich extreme with abundance alternatives of [Fe/H] = −0.79 (Z = 0.003) and [Fe/H] = −0.49 (Z = 0.006) for α = 1.65. The reddening value for 47 Tuc is well determined at E(B−V) = 0.04 ± 0.02. Using the reddening equivalences in BVI given by Taylor (1986) for E(B−V) = 0.04 ± 0.02, we derive E(V−I) = 0.05 ± 0.02 and E(B−I) = 0.09 ± 0.02. The best fit for all of the isochrones shown in Figures 1a-c uses Y = 0.2, [Fe/H] = −0.49, with E(B−V) = 0.04 ± 0.02 and a distance modulus of $(m-M)_V$ = 13.2. In all three color indices we find that the isochrones match the CMDs best for an age of 17 Gyr. The V vs. B−I CMD in Figure 1c shows the good sequence definition provided by the use of the extreme bands as color indices.

Using the horizontal branch magnitude of V_{HB} = 14.05, we find that the absolute magnitude of the red horizontal branch is $M_{V,HB}$ = 0.85 with an uncertainty of around ±0.1. This result clearly suggests that indeed the intrinsic brightness of horizontal branch stars is fainter for metal rich clusters than the canonical value of $M_{V,HB}$ = 0.6.

As for the other clusters now reduced, we note that the main sequence of the nearby and somewhat reddened cluster M4 (E(B−V) ∼ 0.4) has been observed to four magnitudes below the turnoff (Figs. 2a-c), and that the B−I colors range from +1.8 to +3.5. The location of the turnoff itself is clearly defined, as is the narrow sequence of unevolved stars. Comparison of the main sequences of M4 and ω Cen demonstrates the unusual width of the latter (see Figs. 3a-c). This characteristic which results from the relatively large spread in composition occurs not only in the evolved stars above the turnoff, but also in the unevolved members on the main sequence. Thus, it must be concluded that the primordial composition had an unusually wide spread as well.

Table 5 summarizes the basic result for the five clusters now fully reduced. In Figure 4 we have plotted age vs. metallicity for these globular clusters plus the two oldest known open galactic clusters, M67 and NGC 188. We can observe that the ages derived for all of these globular clusters are 17.0 ± 1.5 Gyr, hinting that the globular cluster system might be coeval, and that the epoch of galactic contraction was short. These ages set a lower limit for the age of the universe and thus an upper limit for the Hubble constant of H_o = 58 ± 5 km s^{-1}Mpc^{-1}, assuming q_o = 0.

Table 5
Summary of Basic Results

Cluster	V_{TO}	Color Turnoffs	ΔM_{TO-HB}	[Fe/H] Isochrones	Age (Gyr)
NGC 104[1] (47 Tuc)	17.60 ± 0.10	$B-V = 0.56 \pm 0.02$ $V-I\ = 0.68 \pm 0.02$ $B-I\ = 1.24 \pm 0.02$	3.5 ± 0.2	-0.49	17.0 ± 1.5
NGC 2298[2]	19.50 ± 0.10	$B-V = 0.60 \pm 0.02$ $V-R = 0.27 \pm 0.03$ $R-I\ = 0.39 \pm 0.03$ $V-I\ = 0.71 \pm 0.03$ $B-I\ = 1.30 \pm 0.03$	3.4 ± 0.2	-1.27	17.0 ± 1.5
NGC 5139[3] (ω Cen)	18.30 ± 0.15	$B-V = 0.55 \pm 0.03$ $V-I\ = 0.73 \pm 0.03$ $B-I\ = 1.28 \pm 0.03$	3.8 ± 0.2	-1.27	17.0 ± 1.5
NGC 6121[4] (M4)	16.80 ± 0.07	$B-V = 0.81 \pm 0.02$ $V-I\ = 0.96 \pm 0.02$ $B-I\ = 1.77 \pm 0.02$	3.4 ± 0.2	-1.27	17.0 ± 1.5
NGC 6362[5]	18.75 ± 0.10	$B-V = 0.50 \pm 0.02$ $V-R = 0.31 \pm 0.02$ $R-I\ = 0.35 \pm 0.02$ $V-I\ = 0.68 \pm 0.02$ $B-I\ = 1.18 \pm 0.03$	3.4 ± 0.2	-1.27	16.0 ± 1.5

[1] Alcaino and Liller (1987a) (Results from MIDAS).
[2] Alcaino and Liller (1986b) (Results from IHAP).
[3] Alcaino and Liller (1987b) (Results from MIDAS).
[4] Alcaino and Liller (1988) (Results from MIDAS).
[5] Alcaino and Liller (1986a) (Results from IHAP).

Figure 4. Age versus metallicity for the globular clusters that we
have studied and the two oldest open clusters (NGC 188 and M67).

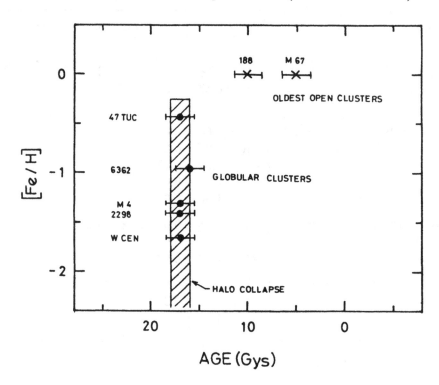

REFERENCES

Alcaino, G. (1973), *Atlas of Galactic Globular Clusters with Colour-Magnitude Diagrams*, (Santiago: Universidad Católica de Chile).

Alcaino, G. and Liller, W. 1986a, *Astron. J.*, **91**, 303.

————. 1986b, *Astron. Astrophys.*, **161**, 61.

————. 1987a, *Astrophys. J.*, **319**, 304.

————. 1987b, *Astron. J.*, **94**, 1585.

————. 1988, *Astrophys. J.*, **330**, 569.

Burstein, D. 1985, *Publ. Astron. Soc. Pacif.*, **97**, 89.

Carney, B.W. 1980, *Astrophys. J. Suppl.*, **42**, 481.

Demarque, P. 1980, in *IAU Symposium No. 85, Star Clusters*, ed. J.E. Hesser, (Dordrecht: Reidel), p. 281.

Harris, W.E. and Racine, R. 1979, *Ann. Rev. Astron. Astrophys.*, **17**, 241.

Harris, W.E., Hesser, J.E., and Atwood, B. 1983a, *Astrophys. J. Lett.*, **268**, L111.

————. 1983b, *Publ. Astron. Soc. Pacif.*, **95**, 951.

Hesser, J.E. and Hartwick, F.D.A. 1977, *Astrophys. J. Suppl.*, **33**, 361.

Lee, S.W. 1977, *Astron. Astrophys. Suppl.*, **27**, 381.

Peterson, C.J. 1986, *Publ. Astron. Soc. Pacif.*, **98**, 1258.

Pilachowski, C.A. 1984, *Astrophys. J.*, **281**, 614.

Sandage, A.R. 1982, *Astrophys. J.*, **252**, 553.

Taylor, B.J. 1986, *Astrophys. J. Suppl.*, **60**, 577.
VandenBerg, D.A. and Bell, R.A. 1985, *Astrophys. J. Suppl.*, **58**, 561.

STELLAR SPECTRUM SYNTHESIS

Jun Jugaku
Tokyo Astronomical Observatory
Mitaka, Tokyo 181
Japan

1 INTRODUCTION

In the ultraviolet region of stellar spectra the absorption lines are numerous and are crowded so that the wings of these lines overlap. Abundance analyses using equivalent widths of stellar spectral lines require the features under investigation to be unblended. However, many of the elements whose abundances are key diagnostics of stellar structure and evolution only have lines in very crowded spectral regions. In such situations the method of spectrum synthesis enables one to obtain information from blended features.

2 HISTORICAL BACKGROUND

Lewis and Liller (1952) were the first to calculate the depression of the continuum level of stellar spectra quantitatively. They chose ε Eri as a test star and computed the profiles of the Ca I $\lambda 4227$ line and the Ca II K line. For the K line they found that reducing the abundance by 30% and raising the continuum by a factor of 1.5 would fit the observations. For the $\lambda 4227$ line they found that raising the continuum by a factor of 1.4 would represent the observations satisfactorily. Since at shorter wavelengths the number of lines per unit wavelength increases compared to that at longer wavelengths, the depression of the continuum level by the overlapping lines must be more effective for the Ca K line region than the $\lambda 4227$ region. Lewis and Liller (1952) correctly concluded that any method which employs the equivalent widths of lines for the analysis of stellar atmospheres must be used with caution; especially so if the lines are chosen from a wide range of wavelengths.

In the early 1960s computers were much less powerful than in recent years. Nevertheless some diligent people made pioneering contributions to the field of spectrum synthesis during this early period. I wish to refer to the works of Climenhaga (1960, 1966). When Bill Liller was at the University of Michigan as a fresh instructor in the Department of Astronomy, both John Climenhaga and I were graduate students in the same office.

3 THE METHOD OF SPECTRUM SYNTHESIS

The atmospheres of carbon stars have low enough temperatures so that molecules are formed and the predominant features in the blue region are absorption bands due to C_2, CN, and CO. It is generally believed that there is more carbon present in these stars than in M-type stars. It is of considerable interest to determine the $^{12}C/^{13}C$ abundance ratio in these stars and to interpret it in terms of the theory of stellar evolution. To ease the tedious computations of radiative transfer for individual lines, Climenhaga (1960, 1966) used the semiempirical equation of Minnaert (1935) to compute the line depths. He paid particular attention to the $\lambda4737$ (1,0) C_2, $\lambda6191$ (4,0) CN, and $\lambda7872$ (2,0) CN bands. Calculations were made for three values of the $^{12}C/^{13}C$ abundance ratio: 100, 20, and 4.

Figure 1. Upper panel: spectrum of the sun and of Arcturus in the wavelength region 4729 Å– 4746 Å. Middle panel: spectra of three R-type stars. Lower panel: Computed molecular profile with an abundance ratio $N(^{12}C^{12}C)/N(^{12}C^{13}C)=2$.

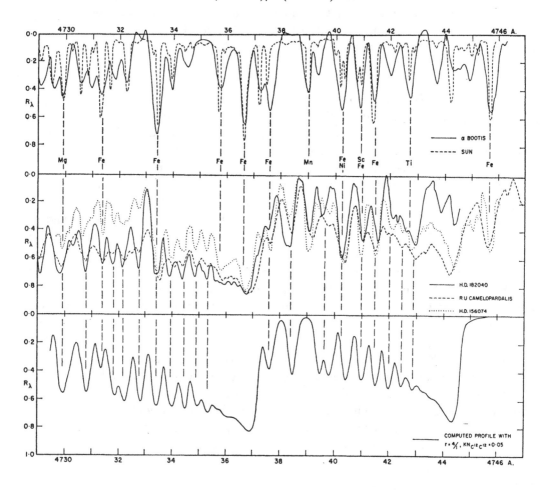

Figure 1 (from Climenhaga 1960) shows a comparison of the spectra of three R-type stars HD 182040, HD 156074, and RU Cam with the spectra of the sun and Arcturus. It also illustrates a calculated profile of the (1,0) $^{12}C^{12}C$ and $^{12}C^{13}C$ molecular bands. Figure 2 (also from Climenhaga 1960) shows a comparison of the computed profiles in the same wavelength region with observed profiles obtained from two spectrograms of the R-type star HD 156074. It appears that the computed profile that would best fit the observed profile would be that for which the abundance ratio $^{12}C/^{13}C$ is somewhat greater than 4. This value is consistent with the equilibrium value in the carbon-nitrogen cycle.

Figure 2. Profiles of bands in the spectrum of HD 156074. The upper panel shows a tracing of a spectrogram obtained with the coudé spectrograph of the 5-m telescope at Palomar, while the lower panel is based on a spectrogram obtained with the three-prism spectrograph of the 1.85-m telescope at Victoria.

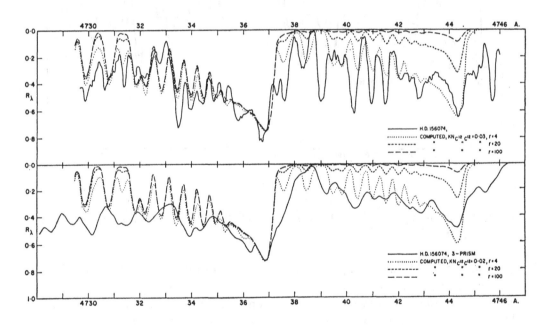

4 RECENT WORK

I obtained my Ph.D. at the University of Michigan in 1957 under the guidance of Professor Lawrence Aller. The title of my thesis was *The Abundance Ratio of Helium to Hydrogen in the Atmospheres of the B Stars*. Bill Liller was one of the members of my thesis committee and I still remember his kind questions in my oral presentation. Since then I have mainly worked on early-type stars. In the past several years, Sadakane, Takada-Hidai, and I have been collaborating in the abundance analyses of normal and peculiar Ap stars. The classical Ap stars are nowadays classified as one group of several families of chemically peculiar (CP)

Figure 3. Computed and observed spectra of Sirius. Top: theoretical spectrum. Bottom: observed spectrum (thick line) compared to theoretical spectrum (thin line) broadened with stellar rotation.

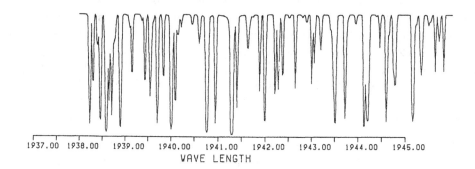

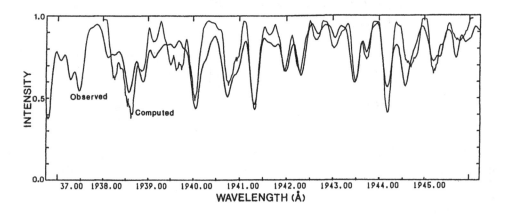

stars populating the interval 7000 K $\leq$ T$_{\text{eff}}$ $\leq$ 30,000 K in overlapping temperature regions (Preston 1974).

Here I wish to present one example of what we have been doing recently with the use of spectral synthesis. The top section of Figure 3 illustrates the computed synthetic spectrum of the wavelength region from 1938 Å to 1946 Å in Sirius. The brightest star in the sky is now recognized as one of the hot Am (CP1) stars. Note the large number of lines with a considerable amount of blending. The line lists and log gf values of Kurucz and Peytremann (1975) and Kurucz (1981) were used in our computations. Synthesized spectra were computed with a modified version of the procedure XLINOP, a subprogram of the ATLAS 6 program (Kurucz 1979). The values of T$_{\text{eff}}$ = 10,000 K, log g = 4.30, and log [N(M)/N(M)$_\odot$] = +0.5 were adopted in the computations of the model atmospheres. Here N(M) denotes the metal abundance of the star and N(M)$_\odot$ that of the sun. In the synthetic spectrum the following abundances were adopted: log [N(Ca,Sc)/N(Ca,Sc)$_\odot$] = -1.0 and log [N(Ti, ..., Ni)/N(Ti, ..., Ni)$_\odot$] = +0.5. We also assumed the classical damping constant in the computations of line profiles. For micro-turbulence we took a value of 2 km/s. Note that a large number of lines are included in the computations with a step width of 0.02 Å. The theoretical spectrum was then broadened with a rotational velocity of v sin i = 18 km/s in the lower part of Figure 3, and compared with the observed spectrum. The thick line was taken from the observations of Rogerson (1987), who recently published a near-ultraviolet spectral atlas of Sirius based on data obtained with the Princeton spectrometer aboard the *Copernicus* satellite.

The most important feature in Figure 3 is the resonance line of Hg II at 1942 Å. There exist 11 isotopic components of the Hg II line in the wavelength region 1942.226 Å– 1942.301 Å. In the present calculations the Hg II line was treated as a blend of these 11 lines and we used the same physical parameters as adopted by Leckrone (1984). The so-called q value (dimensionless mix parameter) was assumed to be terrestrial. After several trials we found that a logarithmic Hg abundance log [N(Hg)/N(H)] = −10.5 gives a good fit of the observed profile. This value is smaller by 0.6 dex than those obtained for normal B stars (Leckrone 1984).

REFERENCES

Climenhaga, J.L. 1960, *Pub. Dominion Astophys. Obs. Victoria*, **11**, 307.

————. 1966, in *Colloquium on Late-Type Stars*, ed. M. Hack, (Trieste: Osservatorio Astronomico di Trieste), p. 54.

Kurucz, R.L. 1979, *Astrophys. J. Suppl.*, **40**, 1.

————. 1981, *Smithsonian Astrophys. Obs. Spec. Rept., No. 390*.

Kurucz, R.L. and Peytremann, E. 1975, *Smithsonian Astrophys. Obs. Spec. Rept., No. 362*.

Leckrone, D.S. 1984, *Astrophys. J.*, **286**, 725.

Lewis, E.M., and Liller, W. 1952, *Astrophys. J.*, **116**, 428.

Minnaert, M. 1935, *Zeit. Astrophys.*, **10**, 40.

Preston, G.W. 1974, *Ann. Rev. Astron. Astrophys.*, **12**, 257.

Rogerson, J.B., Jr. 1987, *Astrophys. J. Suppl.*, **63**, 369.

MASS EXCHANGE AND STELLAR ABUNDANCE ANOMALIES

B.F. Peery, Jr.
Howard University, Washington, DC 20059, USA

Abstract. A brief description of a search for understanding of peculiar-abundance red giants of type MS is presented, illustrated by the example of the MS star HR1105.

INTRODUCTION

Classical stars of type S are variables, their spectra are rich in signatures of elements that are characteristically built up during episodes of neutron irradiation (s process), and absorption-line strengths of the radioactive element, technetium, are dramatic. But there are other red giants with mildly enhanced s-process abundances, and that vary little or not at all. This latter group is particularly noteworthy for its absence of Tc. These stars are often designated type MS.

It is tempting to speculate that MS stars may be cooler analogues of the Ba II stars, which display similar peculiarities. The appeal of such speculation is heightened by the recent observations that Ba II stars are generically spectroscopic binaries, that the few secondaries so far identified are hot subluminous objects, and the suggestion that mass transferred earlier from the evolving secondaries may have contributed the s-processed matter that we now find in the atmospheres of the present primaries (McClure et al. 1980).

Griffin (1984) has shown that the bright MS star HR1105 is indeed a single-line spectroscopic binary, with period 596 days. A hot relic companion of the red giant primary is likely to reveal itself only in the ultraviolet. This report is a brief account of observations of HR1105 with the International Ultraviolet Explorer that do indeed give evidence for the presence of a hot subluminous companion. The observations also suggest indirect clues that may, in other systems, implicate hot companions that are too faint to be observed. More extensive searches for hot companions of MS stars are under way (Ake et al. 1988; Smith & Lambert 1987).

OBSERVATIONS AND QUESTIONS

The search for hot companions of MS-type stars is being carried out with the IUE at low dispersion with the SWP camera. Figure 1 reproduces a spectrogram of HR1105.

The "continuum" level shortward of the C IV emission line
clearly lies above that at longer wavelengths, suggesting
the presence of a hot source of low luminosity. Emission
lines of N V, O III, Si III-IV and C III-IV dominate the
ultraviolet spectrum, adding support to the suggestion of a
hot source in the system. Preliminary evidence indicates
that the emission lines vary in intensity (Ake et al.
1988).

Johnson & Ake (1983) earlier made the serendipitous
discovery of the same high-excitation emission lines in the
spectrum of the MS star HD35155. In Figure 1 an IUE
spectrogram of HD35155 appears below that of HR1105. These
are the two MS stars that have been found thus far to
display these extraordinary ultraviolet spectral features.
Significantly, some Ba II stars also display similar
ultraviolet spectra (e.g., Schindler et al. 1982).

These observations, still meager, keep alive the binary
mass-exchange scenario as the framework for understanding
the peculiar abundances of MS stars. There are taunting
questions: Is the hot source of ultraviolet radiation
an accretion disk, and around which star? Can the
variability of the high-excitation emission lines be

Figure 1. IUE spectrograms of MS stars HR1105
and HD35155 (the latter from Johnson & Ake 1983).

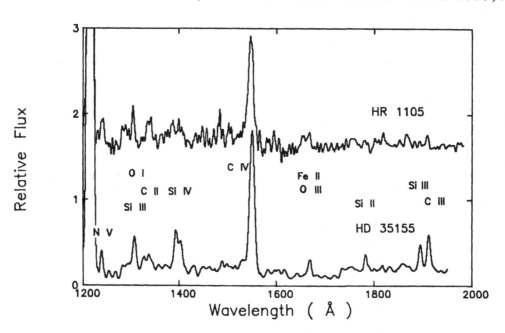

understood as evidence of sporadic episodes of mass transfer? Are the properties of the subluminous companion consistent with the vanishing of Tc since the most recent epoch of neutron-processing? What parameters of the scenario lead to Ba II primaries in some systems and to MS-type primaries in others? Are all MS stars indeed binaries? And from an abundance analysis can we hope to extract clues to the natural history of the remnant's interior?

A SALUTE TO BILL LILLER

I arrived as a graduate student at the University of Michigan without having been prepared for my first sight of a telescope - or of an astronomer. It was inevitable that I should gravitate to Bill Liller as Professor and Friend. First, I was ordered to enroll in his Practical Astronomy course (ocassionally, I still refer to the notes). Second, he loved the music that I was raised on: Jazz!

I cannot detail here my indebtedness to Bill in his role as a mentor and as a human. But permit me a remniscence that is related to this brief report:

Decades later now, I observe with IUE, warm and comfortable and gentlemanly in jacket and tie. Subdued excitement is in the air, like the low noise of the computer fans. More than once I have found myself contrasting this scene with my first heroic attempt to record information from a satellite: Sputnik was in orbit, and Bill and a few of us graduate students were on the roof of Angell Hall in the blustery October night. Our technology was a short-wave broadcast receiver and a wire recorder. I can't claim with certainty that we captured a single Sputnik beep during that passage, but little matter. Bill introduced some new excitement that night that has never quite gone away.

ACKNOWLEDGEMENTS

It is my pleasure to acknowledge the cheerful and expert assistance of the IUE staff in obtaining the data that I have presented here. I could not have reduced the data without the ready advice of the GSFC RDAF staff. This work has been supported in part by NASA grant NAG5-772.

REFERENCES

Ake, T.B., Johnson, H.R. & Peery, B.F. Jr (1988). In A Decade of UV Astronomy with the IAU Satellite. Paris: European Space Agency. (In preparation).
Griffin, R.F. (1984). Observatory,104, 224.

Johnson, H.R. & Ake, T.B. (1983). In Cool Stars, Stellar
          Systems, and the Sun, ed. S.L. Baliunas &
          L. Hartmann, p. 362. New York: Springer-Verlag.
McClure, R.D., Fletcher, J.M., & Nemec, J.M. (1980).
          Astroph. J.,(Letters),238, L35.
Schindler, M., Stencel, R.E., Linsky.J.L., Basri, G.S., &
          Helfand, D.J. (1982). Astroph. J.,263, 269.
Smith, V.E., & Lambert, D.L. (1987). Astron. J. (Preprint).

John Huchra conversing with Nancy Murphy.

THE M31 GLOBULAR CLUSTER SYSTEM [1]

John P. Huchra

Harvard-Smithsonian Center for Astrophysics, 60 Garden St., Cambridge, MA 02138

I. INTRODUCTION

Andromeda's was the first extragalactic globular cluster system discovered. In 1932 Hubble announced the "provisional" identification of 140 nebulous objects around M31 as globular clusters. The objects were resolved on 100-inch plates and fell off in surface number density from the galaxy center. Because Hubble did not know the correct distance to M31 at the time, he thought that these clusters were less luminous than similar globular clusters in our own galaxy.

Further surveys of the extended M31 cluster system were done by Seyfert and Nassau (1945) inspired by Baade, and by Vetesnik (1962). By 1970, there were about 200 suspected clusters. The first detailed spectroscopic and photometric study of the cluster system was done in the late 1960's by van den Bergh (1969) with the Palomar 200-inch Hale Telescope. Van den Bergh found that spectroscopically, the M31 clusters were much like galactic globulars in age but were more metal rich. The metallicity of the clusters did not appear to be a function of radius, but most of his clusters were inside 50 arcmin. Van den Bergh also measured velocities for 33 clusters.

Following the "discovery" of massive galaxy halos by Roberts and Whitehurst (1975, and references therein) from extended HI rotation curves, and by Ostriker and Peebles (1973) and Einasto et al. (1974), Hartwick and Sargent (1974) realized that positions and velocities of globular cluster systems would serve as an excellent independent means for measuring the masses of such halos. Subsequently, two groups, Sargent, Kowal and Hartwick (1977) and Battistini et al. (1980), completed deep and wide area surveys of the M31 system that contain $\sim$ 400 globular cluster candidates.

Other spectroscopic surveys of a very small number of clusters have been done by Spinrad and Schweizer (1972) and by Burstein et al. (1981, 1984). From their photoelectric scanner data, Spinrad and Schweizer confirmed the range of metallicities seen by van den Bergh. They also suggested that three of the M31 clusters were "super metal rich" and that these clusters contained a hot star component, possibly a well populated horizontal branch, unlike similar metal rich galactic globular clusters.

Burstein et al. compared line indices, in particular those from the Hβ, cyanogen and

[1] Work reported here based on observations done with the Multiple Mirror Telescope, a joint facility of the Smithsonian Institution and the University of Arizona.

Mg2 features, measured in M31 and galactic globulars and in elliptical galaxies, and found a small but significant offset in the index versus index plots for the galactic and M31 clusters. Burstein *et al.'s* results confirm the hot star excess seen by Spinrad and Schweizer, but Burstein *et al.* have suggested that this excess is due to the presence of a *young* stellar population, i.e. that star formation last occured in some of the M31 globular clusters only 4 or 5 billion years ago.

The photometric properties of the Andromeda globular clusters have also been investigated in considerable detail since van den Bergh's (1969) work. Photographic photometry has come from the Italian group (Battistini *et al.* 1987; Buonanno *et al.* 1982) and a significant amount of UBV photoelectric photometry has come from the Soviets (e.g. Sharov and Lyutyi 1983, 1985). The major conclusions of the photometric surveys are that the M31 clusters cover the same range in color as the Milky Way's clusters and that the clusters tend to be bluer than the halo at similar radii. A review of the current status of photometric studies can be found in Elson and Walterbos (1988). The photometric work has been of some interest to the question of internal reddening in M31 (Iye and Richter 1985).

There are several good reviews of the general properties of extragalactic globular cluster systems (see *The Harlow Shapley Symposium on Globular Cluster Systems in Galaxies*, IAU Symposium #126, for example). Below I will confine the discussion to my more recent work on the M31 system, work that was inspired by Bill Liller.

II. WORK AT CFA

The Center for Astrophysics entry into the Andromeda globular cluster game came with the identification of over a dozen X-ray point sources with globular clusters in M31 (Van Speybroek, *et al.* 1979). There was, at the time, considerable interest in galactic X-ray sources in globular clusters. The discovery of a population of such sources in another galaxy where they might be studied globally immediately sparked the interest of Bill Liller, who was actively involved in X-ray source identifications as well as detailed studies of galactic globular clusters. Bill felt that I was a good enough spectroscopist and had enough interest in related problems (like cluster system dynamics and the mass of the universe!) to get me involved in the project.

In 1979, we set out to obtain spectroscopy of the X-ray clusters as well as some comparison non-x-ray clusters with the Multiple Mirror Telescope. That telescope had just been commissioned, and Marc Davis, Fred Chaffee and I dragged the "Z-machine," our radial velocity spectrograph from the 1.5-m Tillinghast reflector at Mt. Hopkins, up to the MMT. No luck, lots of clouds.

The next year, a real spectrograph for the MMT was completed and our luck turned. In two good nights, we observed over 60 clusters, almost doubling the number of objects with spectroscopy and velocities. By then, unfortunately, Bill had gone on to more southerly lands. John Stauffer came from Berkeley to help me out, and we wrote up our results and sent them off to the *Ap. J. Letters* (Huchra, Stauffer and van Speybroeck 1982, hereafter HSS). We observed 14 X-ray clusters and 47 others. We found no discernable difference between the spectroscopic properties of the X-ray clusters and the others (not surprisingly in the light of current views on globular x-ray sources!), but we were able to measure the velocity dispersion of the cluster system,

$$\sigma_V \quad = \quad 130 \; km \; s^{-1},$$

and measure the mass of the galaxy halo,

$$M = 3 \; \times \; 10^{11} \; M_\odot \quad (inside \; 30 \; kpc).$$

Perhaps most importantly, we discovered significant rotation of the cluster system — the clusters on the west side of the galaxy had an average positive motion w.r.t. the galaxy center while those on the east were on the average more redshifted. The one-sided velocity difference was

$$V_{ROT} \quad \sim \quad 80 \pm 20 km \; s^{-1}.$$

This was an extremely important result and was both supported by and supportive

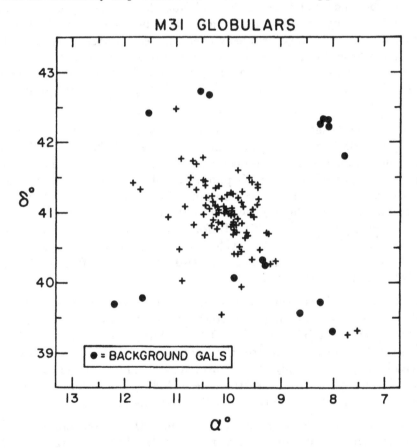

Figure 1. The distribution of observed cluster candidates on the sky. Large filled circles are background galaxies.

of a similar result for the Milky Way globulars. A year earlier, Frenk and White (1980) announced the discovery of a ~ 60 km s^{-1} rotation velocity for the galactic globular cluster system. These discoveries have important implications for the formation and evolution of globular cluster systems since they require either that the clusters were formed from a slowly rotating system or that the clusters (or some of them) have been tidally spun up by interactions with the disk. The latter possibility is considered more likely given the nearly spherical shape of the halo.

It has also been suggested (Freeman 1983, 1985; Zinn 1986) that the globular cluster system is actually composed of two kinematical subsets. There is a metal poor population which is not (or very slowly) rotating and a metal rich population, possibly associated with the galaxy's thick disk, which is rotating and which is probably younger than the metal poor subset. We will discuss this further below.

III. DYNAMICS

Work on the M31 cluster system continued slowly over the last several years. By definition, the next season's observations were completely clouded out — some of you might remember the great flood in Tucson! Nonetheless, mostly as the result of backup observations to other programs, we now have spectra of over 150 M31 globular cluster candidates. We have primarily concentrated on observations of cluster candidates at large projected radii — partly because these objects are the easiest to observe given the background light of M31, but mostly because our primary goals were to confirm the rotation of the cluster system and get a better handle on the metallicity gradient.

Almost all of the new observations have been made with the MMT Spectrograph and a blue sensitive EMI image tube package (affectionately? know as "Big Blue"). This tube has an S20 photocathode so observations can be made to the atmospheric cutoff. Big Blue is coupled to the SAO photon-counting Reticon system (Latham 1982). Because a fiber optic reducing boule has been inserted in the light path, the newer observations have slightly lower resolution, 8-9 Å, than the 5-6 Å resolution of the earler observations (HSS). (Unfortunately this means that, while the new observations provide much better wavelength coverage and thus metallicity measurements, the velocities from single measurements are degraded from ~ 30 km s^{-1} r.m.s. external error to about 45 km s^{-1} r.m.s.

Figure 1 shows the spatial distribution of the objects observed. The total number of actual clusters is 127. The remaining ~ 25 candidates are mostly background galaxies or, in a few cases, compact HII regions or OB associations mistakenly identified as globulars. As can be seen in the figure, the fraction of background galaxies among the cluster candidates rises steeply with projected radius; most of the candidates at large radii are in fact galaxies. Unlike the case of M87 (Huchra and Brodie 1984), the galaxies behind M31 are not mostly in a single galaxy cluster.

Figures 2 and 3 show the velocity histogram for the M31 globular clusters and the velocity as a function of projected radius, respectively. Following van den Bergh (1969), the projected radius is defined as

$$R_p = (x^2 + y^2)^{1/2}$$

where x and y are the projected distance in arc minutes from the minor and major axes. Neither plot is surprising; the velocity histogram is consistent with a Gaussian and its mean is -281 ± 25 km s^{-1}, in extremely good agreement with the heliocentric velocity of the galaxy, -297 ± 1 km s^{-1}, determined from the nucleus and from global 21-cm neutral hydrogen observations. The l.o.s. velocity dispersion stays fairly constant in the inner parts of the galaxy (inside 1 degree), but may be dropping outside that. Figure 3 does not have the effects of rotation removed.

A simple dynamical analysis can be done using either the virial or projected mass estimators (Bahcall and Tremaine 1981; Heisler, Bachall and Tremaine 1985; see Huchra and Brodie, 1987 for an example). We assume that the distance of M31 is 750 kpc. The measured line-of-sight velocity dispersion, $\sigma_V = 150$ km s^{-1}. Both estimators give the galaxy's mass and thus mass to light ratio as

$$M \quad \sim \quad 3.2 \pm 0.5 \times 10^{11} M_\odot$$

and

$$M/L \quad \sim \quad 30 \pm 5 \quad M_\odot/L_\odot$$

as before, in very good agreement with the 21-cm rotation curve mass.

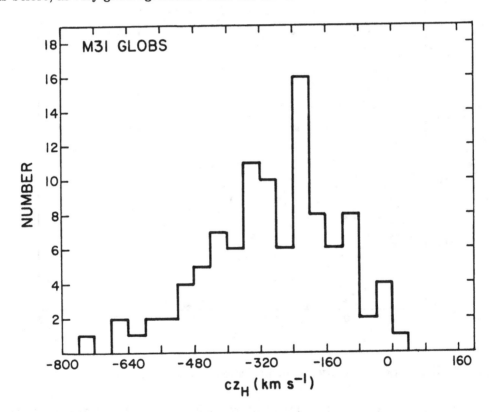

Figure 2. The velocity histogram for the M31 globular clusters. Note that all velocities quoted in this paper are heliocentric.

IV. METALLICITY

The integrated spectra of old stellar populations contain a large number of features that are sensitive to chemical abundance. Most early metallicity work concentrated on just a few of these, such as the strength of the CaII H and K lines or the associated break in the ultraviolet continuum (e.g. van den Bergh 1969; Zinn 1980), to estimate a mean [Fe/H]. As part of a large collaborative effort to study the spectroscopic properties of extragalactic globular cluster systems, in the last few years we have developed a reasonable calibration of a larger variety of line indices to [Fe/H]. Figure 4 shows a portion of the spectrum of one of our MMT template galaxies, taken with the same instrumental setup used for the globular cluster observations.

Esssentially, such calibrations are derived by plotting measured line indices versus [Fe/H] for objects of known metallicity. Line indices are defined in magnitudes as

$$I = -2.5 \, log \left[\frac{2F_I}{(F_{C1} + F_{C2})} \right]$$

where F_{C1} and F_{C2} are the mean fluxes in continuum bandpasses on either side of

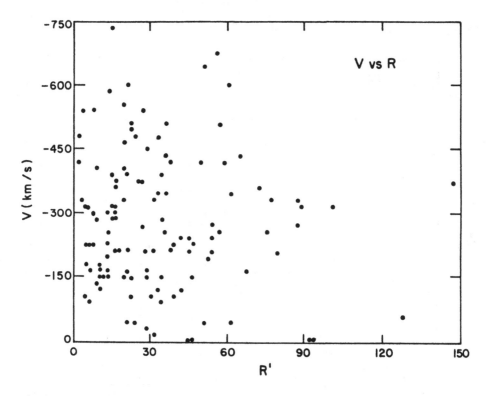

Figure 3. Heliocentric velocity versus projected radius, R_P, in arc minutes.

the feature, and F_I the mean flux in the feature itself. Those indices which are simple colors are defined as

$$I = -2.5 \, log \, (F_2/F_1) \quad ,$$

where the definitions of F_2 and F_1 are obvious.

Figure 5 shows an example of the calibration of one of the best metallicity indicators, the Mg2 feature at 5190 Å. For this index, the feature bandpass is 5156.00–5197.25 Å, and the continuum bandpasses are 4897.00–4958.25 Å and 5303.00–5366.75 Å.

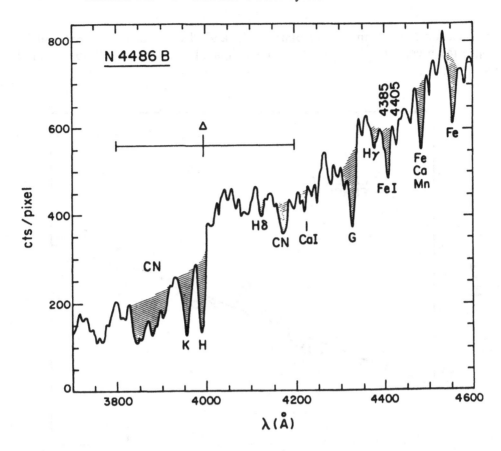

Figure 4. Blue spectrum of the template object NGC4486B with various features useful for abundance determinations marked and/or cross-hatched.

Table 1. [Fe/H] CALIBRATION

Best Indices		λ Å	R
	Δ	4000	0.90
	Mg2	5177	0.92
	CH	4300	0.90
Next Best			
	CN_B	3860	0.83
	HK	3960	0.80
	CN_R	4161	0.77
	MgH	5153	0.84

Table 1 lists the 7 indices with the best correlation with [Fe/H]. Also given is their approximate central wavelength and the linear correlation coefficient of the fit of index to [Fe/H]. The Δ index is an intermediate band color defined by Brodie and Hanes (1986) which measures the H+K break. The MgH index measures the strength of the magnesium hydride molecular band, CN_R and CN_B measure the red and blue cyanogen features, and CH is the G-band which is a blend of iron, CH and other elements.

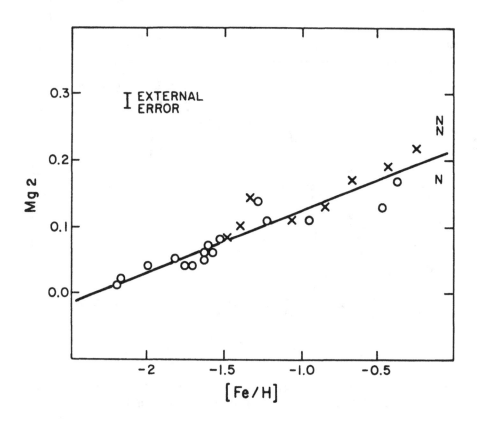

Figure 5. The calibration of the Burstein-Faber Mg2 index versus mean metallicity [Fe/H]. The X's are M31 clusters with metallicities determined from IR colors, the O's are galactic globular clusters, and the N's are giant stars in NGC 188 (with an assumed solar [Fe/H]).

The calibrations discussed above are based on the metallicity scale of Zinn (1980) and the cross-calibration of infrared colors on metallicity used by Frogel, Persson and Cohen (1980). These will soon be updated to the Zinn and West (1984) scale. The details of the calibration work, including the definitions of the bandpasses for various features and a discussion of the errors will be given in Brodie and Huchra (1988) and Huchra, Brodie and Stauffer (1988).

Figure 6 graphically presents the results for the Andromeda clusters. The mean metallicity for our 127 clusters in M31 is [Fe/H] $\sim$ -1.2. This is slightly more metal rich than the Milky Way's clusters where the mean [Fe/H] $\sim$ -1.4. There is a measurable, but weak metallicity gradient; in particular, the scatter in [Fe/H] is large at small projected radii and decreases with R. Even though the upper envelope of [Fe/H] appears to be decreasing rapidly with radius, the mean actually appears to decrease much less rapidly. More cluster observations are needed at radii greater than 1 degree before any definitve mean gradient can be measured.

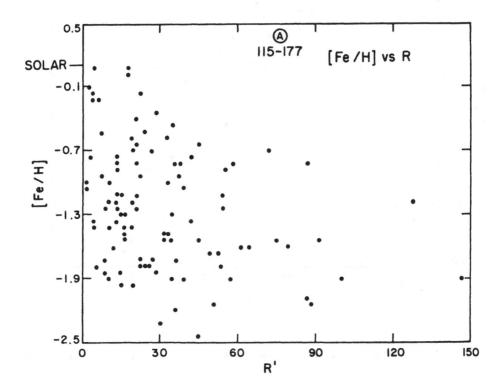

Figure 6. Metallicity, [Fe/H], versus projected radius, R_P, in arc minutes. The unusual cluster, 115-177, is marked

The most metal rich clusters are about solar abundance. In fact, 115-177, the strongest lined cluster in our sample, is MORE than solar. This cluster candidate has an extremely strange spectrum including weak [OIII] 5007Å emission, which may indicate the presence of a planetary nebula (Huchra and Burstein, in preparation).

V. ROTATION OF THE M31 GLOBULARS

Since the discovery of the rotation of the Andromeda cluster system (HSS) and the Milky Way (Frenk and White 1980), a small controversy has arisen over whether or not these globular cluster systems can be broken into distinct kinematic subsamples. Although the situation for the Galaxy was initially somewhat muddled (Frenk and White 1982), most recent analyses have tended to confirm the dual nature of the cluster kinematical population (Zinn 1985,1986). Essentially, the metal poor clusters ([Fe/H] $\leq$ -0.6) form a slowly rotating ($V_r \sim$ 60 km s^{-1}), spheroidal population, while the metal rich systems are more disklike as well as much more rapidly rotating ($V_r \sim$ 150 km s^{-1}).

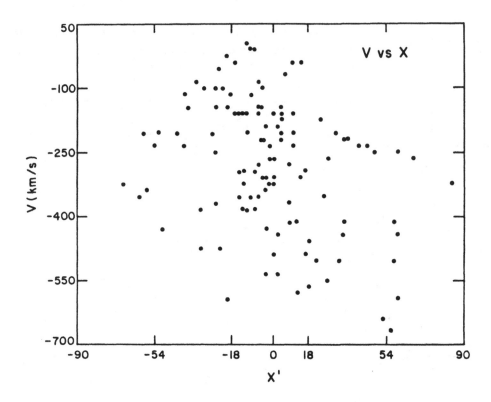

Figure 7. Velocity versus x, the projected distance from the minor axis, in arc minutes.

A similar result has been quoted by Freeman (1983, 1985) and Searle (1983) for the Andromeda system based on our old (HSS) velocities and spectra for 61 clusters. Dividing the M31 clusters again at [Fe/H] = -0.6, Freeman states that the metal poor clusters are not rotating at all and have a velocity dispersion of only 90 km s^{-1}. He claims that the metal rich clusters essentially follow the disk population I rotation curve. Searle, using his own unpublished sample of 100 clusters, no longer sees any evidence of a separate disk population of M31 globular clusters (1986 as quoted in Zinn 1986).

The current state of our observations of the cluster system rotation can be seen in Figure 7 and is summarized in Table 2. The system rotation is evident in Figure 7, where the clusters at negative x have more positive velocities than the mean and the clusters at positive x average more negative. The amplitude of the rotation is unchanged from HSS, 156 km s^{-1} peak-to-peak or 78 ± 19 km s^{-1}.

We can also cut the sample on metallicity. Table 2 presents the results when the clusters are split roughly about the mean metallicity. V+ is the mean velocity for the clusters in the subsample with x ≥ 15', and V- is the mean velocity for the clusters with x ≤ -15'. ΔV is the difference between the subsample means. For both the metal rich and metal poor subsamples there are significant differences between positive and negative sides. The differences between the metal rich and poor subsamples is

Table 2. Kinematics of Clusters at Large Radii

[Fe/H] greater than -1.2		(66)
	V+ = -378 ± 34	(14)
	V- = -237 ± 43	(16)
	ΔV = 141 km s^{-1},	mean = -287 ± 12 km s^{-1}
[Fe/H] less than -1.2		(44)
	V+ = -388 ± 44	(14)
	V- = -230 ± 48	(9)
	ΔV = 158 km s^{-1},	mean = -289 ± 43 km s^{-1}
For ALL clusters		(127)
	V+ = -379 ± 27	(31)
	V- = -223 ± 27	(31)
	ΔV = 156 km s^{-1},	mean = -281 ± 25 km s^{-1}

For clusters with projected minor axis distance, x $\geq$ 15'.
Number in parentheses the number of clusters in the sample.
Error quoted is the r.m.s. error of the mean.

statistically insignificant. The system rotation is detected at better than 4 sigma.

Contrary to popular opinion, our data do not support the hypothesis that the cluster rotation is a strong function of [Fe/H]. It is possible that different cuts in [Fe/H] might produce significant differences between the metal poor and metal rich cluster kinematics; a quick look at our data does not show any greater differential between metal rich and metal poor when the cut is made at [Fe/H] = -0.6. A better determination will be attempted when our new metallicity calibration is completed. It does appear from our data, however, that the metal poor clusters are rotating. Unfortunately it is not yet possible to completely deproject the M31 clusters.

VI. SUMMARY

In many ways, because it is near, yet not too near, the M31 cluster system is the best studied of all globular cluster systems. The current status of our observations of the M31 system is unfortunately simple to describe. We've found no particular surprises. The mass derived from the velocity observations is consistent with mass estimates from other techniques — which is reassuring. The blue mass-to-light ratio is typical of massive spirals, $\sim$ 30 in solar units, and is both above the M/L predicted from simple stellar population models of the galaxy's bulge + disk and also well below the "critical" M/L which would be required if the mass in galaxies was to close the universe (about 2400 in solar units for a Hubble constant of 100 km s^{-1}). As usual, if the M/L of M31 is typical, Ω_G, the ratio of the observed mass density in galaxies to the closure mass density, is only $\sim$ 1%.

The average metallicity, [Fe/H], for the Andromeda clusters is slightly higher than that of the Milky Way's clusters. This was also "predicted" by the empirical correlation that seems to exist between a galaxy's luminosity or mass and its metal abundance. M31 is more massive than the Galaxy. The line-strength/luminosity relation has been commented on often before, and is probably crucial in understanding the formation of galaxies. The evidence in hand for several globular cluster systems (see Huchra, 1988, for a summary) indicates that a similar relationship between [Fe/H] and parent galaxy luminosity holds for those systems as well.

The M31 clusters exhibit a large range in [Fe/H], with the most metal rich having approximately solar abundances. It appears that there is some correlation of cluster [Fe/H] with radius. Whether this is best described as a gradient, a decrease in the upper "envelope" of abundance, or just a statistical (population) effect is difficult to say without more data and a more detailed analysis. We also partially confirm the results of Burstein *et al.* (1981, 1984) — there is a weak Hβ line anomaly in the M31 clusters w.r.t the galactic clusters — but we find the amplitude of the mean difference in the Hβ versus Mg2 plot to be less than they reported.

The $\sim$ 80 km s^{-1} rotation of the cluster system is confirmed at better than 4 sigma, but we have not found a signifcant difference in the rotation properties of the metal rich clusters w.r.t. the metal poor clusters. Whether this also represents a difference between M31 and the Milky Way remains to be seen. By definition, we know we can get better velocities, better metallicities and a much larger sample of objects to work with — there are still over 250 Andromeda clusters waiting to be observed. If only Bill were back in the Northern Hemisphere!

I would like to give my special thanks to Bill Liller who snookered me into the globular cluster game almost ten years ago. I would also like to thank my colleagues in this work, John Stauffer and Jean Brodie, and Susan Tokarz, who has helped immensely with the data reduction and table entry. This work has been supported by the Smithsonian Institution.

REFERENCES

Bahcall, J. and Tremaine, S. 1981, *Ap. J.* **244**, 805.

Battistini, P., Bonoli, F., Braccesi, A., Fusi Pecci, F., Malagnini, M., and Marano, B. 1980, *Astron. and Ap. Suppl.* **42**, 357.

Battistini, P., Bonoli, F., Braccesi, A., Federici, L., Fusi Pecci, F., Marano, B. and Borngen, F. 1987, *Astron. and Ap. Suppl.* **67**, 447.

Brodie, J. and Hanes, D. 1986, *Ap. J.* **300**, 258.

Brodie, J. and Huchra, J. 1988, *Ap. J.* in press.

Buonannao, R., Corsi, C., Battistini, P., Bonoli, F. and Fusi Pecci, F. 1982, *Astron. and Ap. Suppl.* **47**, 451.

Burstein, D., Faber, S., Gaskell, M. and Krumm, N. 1981, in *I.A.U. Colloquium 68: Astrophysical Parameters for Globular Cluster*, A. G. D. Phillip and D. S. Hayes, eds. (Schenectady: L. Davis Press), p441.

Burstein, D., Faber, S., Gaskell, M. and Krumm, N. 1984, *Ap. J.* **287** 586 (BFGK).

Einasto, J., Kaasik, A. and Saar, E. 1974, *Nature* **250**, 309.

Elson, R. and Walterbos, R. 1988, *Ap. J.* in press.

Faber, S. 1973, *Ap. J.* **179**, 731.

Faber, S., Friel, E., Burstein, D. and Gaskell, M. 1985, *Ap. J. Suppl.* **57**, 711.

Freeman, K. 1983, in *Internal Kinematics and Dynamics of Galaxies* , I.A.U. Symposium # 100, E. Athanassoula, ed., (Dordrecht: Reidel), p359.

Freeman, K. 1985, in *The Milky Way Galaxy*, I.A.U. Symposium # 106, ed. H. van Woerden, (Dordrecht: Reidel).

Frenk, C. and White, S. 1980, *M. N. R. A. S.* **193**, 295.

Frenk, C. and White, S. 1982, *M. N. R. A. S.* **198**, 173.

Frogel, J., Persson, E. and Cohen, J. 1980, *Ap. J.* **240**, 785 (FPC).

Hartwick, F. D. A. and Sargent, W. L. W. 1974, *Ap. J.* **190**, 283.

Heisler, J., Tremaine, S. and Bahcall, J. 1985, *Ap. J.* **298**, 8.

Hubble, E. 1932, *Ap. J.* **76**, 44.

Huchra, J. 1988, in *The Harlow Shapley Symposium on Globular Cluster Systems in Galaxies*, I.A.U. Symposium #126, J. Grindlay and A. G. D. Philip, eds. (Dordrecht: Reidel), p255.

Huchra, J. and Brodie, J. 1984, *Ap. J.* **280**, 547.

Huchra, J. and Brodie, J. 1987, *A. J.* **93**, 779.

Huchra, J., Brodie, J. and Stauffer, J. 1988, in preparation.

Huchra, J., Stauffer, J. and van Speybroeck, L. 1982, *Ap. J. Letters* **259**, L57 (HSS).

Iye, M. and Richter, O.-G. 1985, *Astron. and Ap.* **144**, 471.

Latham, D. 1982, in *Instrumentation for Astronomy with Large Optical Telescopes*, ed. C. M. Humphries (Dordrecht: Reidel), p. 259.

Ostriker, J. and Peebles, P. J. E. 1973, *Ap. J.* **186**, 467.

Roberts, M. and Whitehurst, R. 1975, *Ap. J.* **201**, 327.

Sargent, W. L. W., Kowal, C., Hartwick, F. D. A., and van den Bergh, S. 1977, *A. J.* **82**, 947.

Searle, L. 1983, private communication.

Seyfert, C. and Nassau, J. 1945, *Ap. J.* **102**, 377.

Sharov, A. and Lyutyi, V. 1983, *Astrp. and Sp. Sci* **90**, 371.

Sharov, A. and Lyutyi, V. 1985, *Sov. Astron. Lett.* **11**, 248.

Spinrad, H. and Schweizer, F. 1972, *Ap. J.* **171**, 403.

van den Bergh, S. 1969, *Ap. J. Suppl.* **19**, 145.

van Speybroek, L., Epstein, A. Foreman, W., Giacconi, R., Jones, C., Liller, W. and Smarr, L. 1979, *Ap. J. Letters* **234**, L45.

Vetesnik, M. 1962, *Bull. Astron. Inst. Czech.* **13**, 180.

Zinn, R. 1980, *Ap. J. Suppl.* **42**, 19.

Zinn, R. 1985, *Ap. J.* **293**, 424.

Zinn, R. 1986, in *Stellar Populations*, C. Norman, A. Renzini and M. Tosi, eds. (Cambridge: Cambridge University Press), p73.

Zinn, R. and West, M. 1984, *Ap. J. Suppl.* **55**, 45.

DISCUSSION

L. Robinson: From the metallicity gradient, can you tell how long M31 took to form its globular clusters?

J. Huchra: I'd love to, but I can't. The models for chemical enrichment versus radius are soft enough such that timescales between 10^6 and 10^9 years could be fit.

J. Grindlay: Could you comment further on the limits that can be set on the possible correlation between [Fe/H] and rotational velocity? It would be very interesting to determine whether the M31 globulars that are most metal-rich are concentrated in a (thick) disk as in our own galaxy.

J. Huchra: The metal-rich globulars have essentially the same (141 versus 158 km s^{-1}) peak-to-peak "rotational" difference but the errors on these numbers are large. We have also not examined the consequences of other choices of the break between high and low metallicity. The statement that had been made by Searle, Freeman and others that the low metallicity globulars are **not** rotating is, however, clearly wrong.

S. Strom: Can you compare the metallicity of the globulars with M31's halo (or that of other Sb's)? Also, can you compare the velocity dispersion profile of the cluster system with that of the halo?

J. Huchra: The only two galaxies (Sb's) with spectroscopic profile measures (for the globular clusters) are our own and M31. We've tried working on M81 and have not yet succeeded. All I can say is that the metallicity in both the halo and globulars seems to fall with radius. There is some evidence in M87 (not an Sb) from Judy Cohen that the globulars and halo have different metallicities at the same radii, just as you determined. As far as velocity dispersions, there is agreement at the core of M31 between the clusters and the stars, and the bulge dispersion is also falling with radius, but it is not really measured very far out.

Debbie Elmegreen (left) conversing with Francis Wright.

SPIRAL STRUCTURE AND STAR FORMATION IN GALAXIES

Debra Meloy Elmegreen
IBM Watson Research Center, Yorktown Heights, NY
and Vassar College Observatory, Poughkeepsie, NY

INTRODUCTION

In 1978 I approached Bill Liller about doing a thesis on dust in spiral galaxies. The project required surface photometry data over a wide wavelength range, from ultraviolet through near-infrared, so photographic plates were the preferred medium. Bill's expertise in optical observations made him the logical advisor, since I had only done millimeter wave observations up to that point and needed to learn how to become an optical astronomer. Bill accompanied me to what was then Agassiz Observatory on many a cold night (including one with a -10° chill factor!), and taught me observational techniques ranging from darkroom setup to standard star photometry. His key lesson was an introduction to the use and hypersensitization of IV-N emulsions; I still use them in much of my current research on the near-infrared structure of spiral galaxies. Bill made a few phone calls and arranged for a Palomar Schmidt run so that I could start serious observing. Next, he assisted me in writing my first observing proposal to Cerro Tololo, where I had a month-long run on the Curtis-Schmidt, the 1.5m, and the 4m telescopes. I had perfect weather and obtained all my thesis data then. When I returned, Bill instructed me on the use of the microdensitometer so that the plates could be scanned and calibrated.

For my thesis, I estimated the densities of dust lanes and dusty regions in nearby spiral galaxies from a radiative transfer analysis of multicolor surface photometry data (Elmegreen 1980). Among other things, I was seeking the optical counterparts to the giant molecular clouds that had recently been discovered through carbon monoxide observations. Some of the dusty regions in my program galaxies were found to be extremely massive. Extensive clouds and their relation to large scale star forming regions in galaxies continue to be among my primary areas of research.

One of the galaxies studied in my thesis was NGC 7793; I examined other more conventional spirals as well, but this galaxy was a fortuitous choice for many reasons. The rather chaotic spiral appearance of NGC 7793, compared with the more orderly spiral arms in familiar galaxies like NGC 5194, led me to consider the details of spiral morphology and the relation between arm structure, internal properties, and external perturbations in a wide range of spiral Hubble types. This has been another of my main research interests since my Harvard days.

Bill's instructions on the use of IV-N emulsions undoubtedly helped me to become a Carnegie Postdoctoral Fellow, for my proposed observing project at Hale (now The Mount Wilson and Las Campanas Observatories) was a study of the structure of spiral galaxies in the blue and near-infrared passbands. I had 100 nights observing per year for two years, with 65% clear weather, so I amassed stockpiles of data. It was ironic that Hale had no microdensitometer (Caltech was in the process of building one), so I postponed surface photometry till I went to IBM after my fellowship ended. Amazingly enough, IBM does have a microdensitometer; it belonged to the x-ray department at Watson Research Lab and was modified for astronomical plate scanning (Angilello et al. 1984). Despite its slowness, (though no slower than Harvard's microdensitometer that I used for my thesis!), we have now essentially completed scanning the hundred or so galaxies photographed in different passbands.

LARGE SCALE REGIONS OF STAR FORMATION

Properties of large scale regions of star formation were first studied optically by investigating the distributions of OB stars and HII regions in the Milky Way (Morgan et al. 1952) and in external galaxies (Baade 1963). More recently, aperture synthesis radio observations of external galaxies have revealed massive extended HI clouds (e.g., Allen et al. 1973, Newton 1980a,b). We compared the properties of well-ordered "beads on a string" HII regions with radio observations in two dozen galaxies like NGC 5248 (Elmegreen & Elmegreen 1982a), shown in Figure 1. The regular positioning of the giant HII regions in such galaxies made them ideal for studying properties related to global star forming regions. We found that the spacing between giant HII regions correlated with the optical size (R_{25}) of the galaxy, and that the HII regions had spacings similar to those between concentrations of HI emission seen in 21cm aperture synthesis observations. Such HI clouds, or "superclouds", typically contain on the order of $10^7 M_o$, extending over several hundred parsecs, and the dense cores within them form stars. Superassociations and star complexes are the visual manifestations of coherent star forming activity within the superclouds. Gould's Belt in the Milky Way appears to be a local example of one such region (Elmegreen 1985).

Superclouds such as those found in external galaxies are evident in the HI contours of McGee and Milton (1964) and in the Henderson, Jackson and Kerr (1982) survey of the outer parts of the Milky Way galaxy. We sought superclouds in the inner Galaxy by using the Weaver-Williams 21cm survey of the northern hemisphere. Thirty-eight superclouds were detected in the first galactic quadrant, concentrated predominantly in the Sagittarius and Scutum spiral arms (Elmegreen & Elmegreen 1987). Most were associated with giant molecular clouds. The masses of the inner Galaxy superclouds were determined to be several times $10^7 M_o$; all together, they contain 60% of the gas mass in the first galactic quadrant. The cloud densities exceed the critical densities for gravitational   binding, and the velocity dispersions are comparable to the virial theorem linewidths. The emerging scenario is one in which galaxies form superclouds as the largest cloudy units, which encompass giant molecular clouds; hierarchical subfragmentation then leads to star formation in the densest cores.

Currently I am studying details of star formation in the nearby supercloud encompassing the M17 giant molecular cloud by searching for those stars which are most likely to be embedded early main sequence or pre-main sequence stars (Elmegreen et al. 1987). B and I plates were blinked to reveal 2100 reddened stars in a 4 degree by 5 degree area, and color-magnitude diagrams were produced to estimate the possible spectral types, extinctions, and distances. Coordinates were derived for comparison with the IRAS point source catalog and the CO contours of the molecular cloud; we have made over 50 optical identifications for IRAS point sources within the cloud hotspots. IRAS 12-25-60 micron flux data imply that an additional 70 sources within the M17 molecular cloud are deeply embedded low mass objects, possibly T Tauri stars or protostars.

Figure 1: A computer image of NGC 5248 photographed in the ultraviolet passband with the Palomar 5m telescope is shown. Note the prominent HII regions beading the spiral arms.

SPIRAL STRUCTURE

My investigations of spiral arm structure have concentrated on distinguishing between galaxies containing spiral density waves and those that probably have none. The different arm structures reflect the internal and external perturbations that created them, and those processes can be better understood by examining density enhancements of different stellar populations as highlighted by different passbands. At Palomar, a wide range of spiral Hubble types was photographed in order to examine the details of variation in spiral arm structure (Elmegreen 1981). We developed an arm classification system which depends on the arm symmetry and degree of continuity (Elmegreen & Elmegreen 1982b, 1987). At one extreme are galaxies which have a classic "grand design" spiral structure with long symmetric arms, such as NGC 5194 (Arm Class 12); at the other extreme are galaxies that we call "flocculent" spirals because of the patchy spiral arms, such as NGC 7793 (Arm Class 2), shown in Figure 2. The blue and near-infrared images of grand design spirals are remarkably similar; both show strong and well defined spiral arms. Flocculent galaxies have very different B and I appearances. The blue image shows some bright OB associations and HII regions, whereas the near-infrared photograph, which highlights light from slightly older stars, is much more homogeneous.

Figure 2: Shown is the computer image of the flocculent spiral NGC 7793 photographed in the U band with the Cerro Tololo 4m telescope.

A quantitative analysis of spiral structure in 50 barred and nonbarred galaxies was made by examining azimuthal and radial scans of deprojected B and I images (Elmegreen & Elmegreen 1984, 1985). In flocculent spiral galaxies such as NGC 5055, several blue peaks are seen in azimuthal scans, but the contrast in the near-infrared is much less prominent. On the other hand, in grand design spirals such as NGC 4321, both the blue and the near-infrared scans show strong arm-interarm contrasts. These quantitative measurements confirmed our qualitative assessments; the grand design arms are strong in both passbands, but in the flocculent arms only asymmetric blue peaks appear. Spiral arm class is correlated with density waves: grand design galaxies evidently have strong spiral density waves, whereas flocculent galaxies have very weak or nonexistent waves. Flocculent galaxies may derive their morphology solely from sheared chains of propagating star formation (Seiden and Gerola 1982).

We found that the range of spiral arm morphology in barred galaxies is similar to that in nonbarred galaxies. In addition, we discovered two distinct types of bars: galaxies like NGC 1300 have a constant bar surface brightness with radius, so we call them "flat-barred" galaxies. Those like NGC 1073, in which the bar surface brightness decreases with radius, are called "exponential barred" (Elmegreen & Elmegreen 1985). A number of correlations with size, rotation curves and spiral arm morphology implies that the two types of bars are formed by different mechanisms.

We applied our arm classification system to all 708 northern hemisphere galaxies on the Palomar Sky Survey with diameters larger than 1 arcminute and inclinations less than 55° (Elmegreen & Elmegreen 1982, 1987). Statistical analyses of the effect of environment on spiral arm class revealed that, among all nonbarred spiral galaxies, the fraction that have a grand design rapidly increases with group crossing rate until reaching a saturated value of 66% (Elmegreen & Elmegreen 1983, 1987). The statistics are similar for barred spirals in groups, unless a barred spiral is in a binary configuration. In such cases, 90% of the spirals have a grand design. The interesting exceptions are the 10% which have both an internal driving mechanism from a bar and an external perturbation from a companion, but which still have a flocculent structure. The linear sizes of all flocculent galaxies are statistically smaller than those of grand design galaxies by a factor of 50%, so that our angular size-limited sample of classified galaxies encompasses a larger volume for grand design than for flocculent galaxies. Correcting for this volume factor of 3.3, we find that two-thirds of all spiral galaxies have a flocculent structure.

Currently we are producing color enhanced images of approximately 200 spirals as a method for examining spiral arm structure (Seiden et al. 1987). Rectification reveals spiral structure in a dramatic way. The continuity of spiral arms may be traced more easily than in conventional photographs, and in some cases is seen to extend more than 540°. Inner ovals have been discovered in some galaxies. Star formation patterns such as chains and superassociations are apparent at a glance in the composite color photos. For approximately 100 galaxies, photographic images have been made in the B and I passbands. For about 10% of these, images in U, V, and R passbands were also taken. An additional 100 galaxies have been photographed in the v, g, and r passbands using a direct-SIT vidicon camera. The images are constructed by scanning and calibrating the B and I plates, normalizing and aligning them to one another, averaging them to produce a fake G image, and co-adding the three images (or the vgr images). The color image is balanced so that the bright star forming regions appear bluish and the the spiral arms are white. This color balancing allows easy identification of the star forming regions (a true color photograph of a galaxy would appear mostly yellow since the middle spectral types dominate, and the blueness of the arms is only a slight enhancement over the color of the interarm region). The color image is deprojected by stretching along the minor axis for a clear view of the spiral structure. An azimuthally averaged radial profile is subtracted so that both the inner and outer regions of a galaxy can be viewed at once; the effective dynamic range is thus increased. Figure 3 shows the rectified I band computer image of the grand design spiral NGC 598 with the radial profile subtracted.

Figure 3: A computer image of NGC 598 photographed in the I band with the Palomar 1.2m Schmidt telescope is shown. The image has been deprojected by stretching it along the minor axis, and the averaged radial profile has been removed. The bar on the left indicates 100 pixels, while the bar on the right indicates R_{25}. Note the symmetric appearance of the inner arms in this grand design galaxy. The kinking of the spiral arms in the central regions is an indication that a central oval is present.

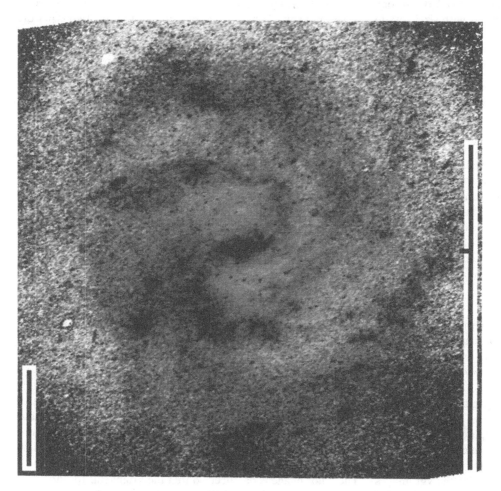

I have now come full circle since my thesis, by participating in a collaborative effort to examine details of the grand design galaxy NGC 3031 in several wavelengths (Kaufman *et al.* 1987). My role is to determine dust lane positions and extinctions so that they may be compared with positions and strengths of predicted and observed shocks. Thus, for the first time since my thesis was completed, I am again doing radiative transfer analyses.

Much work remains in observations of spiral structure in order to sort out such effects as star formation efficiencies, metallicities, initial mass functions, rotation curves, and massive halos on the morphology. I owe Bill Liller a debt of gratitude for giving me my start on the optical observations of galaxies which I enjoy so much.

REFERENCES

Allen, R.J., Goss, W.M., van Woerden, H. (1973), Astron.Astrophys., 29, 447.

Angilello, J., Chiang, W.-H., Elmegreen, D.M., and Segmuller, A. (1984), NASA Conference on Astronomical Microdensitometry, ed. D. Klinglesmith, NASA Publication no. 2317, 229.

Baade, W. (1963). The Evolution of Stars and Galaxies, ed. C. Payne-Gaposchkin, chpt. 16. Cambridge, Harvard University Press.

Elmegreen, B.G. (1985). Physica Scripta, vol. T11, 48.

Elmegreen, B.G. and Elmegreen, D.M. (1982b). Monthly Not.R.Astron.Soc., 203, 31.

Elmegreen, B.G. and Elmegreen, D.M. (1983). Astrophys.J., 267, 31.

Elmegreen, B.G. and Elmegreen, D.M. (1985). Astrophys.J., 288, 438.

Elmegreen, D.M. (1980). Astrophys.J.Suppl., 43, 37.

Elmegreen, D.M. (1981). Astrophys.J.Suppl., 47, 229.

Elmegreen, D.M., Beck, K., Howard, J., Phillips, J. and Thomas, H. (1987). in preparation.

Elmegreen, D.M. and Elmegreen, B.G. (1982a). Monthly Not. R. Astr. Soc., 201, 1021.

Elmegreen, D.M. and Elmegreen, B.G. (1984). Astrophys.J.Suppl., 54, 127.

Elmegreen, D.M. and Elmegreen, B.G. (1987). Astrophys.J., 314, 3.

Henderson, A.P., Jackson, P.D., and Kerr, F.J. (1982). Astrophys.J., 263, 116.

Kaufman, M., Bash, F., Hodge, P., Kennicutt, R. and Elmegreen, D.M. (1987). in preparation.

McGee, R.X. and Milton, J.A. (1964). Austral. Phys., 17, 128.

Morgan, W.W., Sharpless, S. and Osterbrock, D. (1952). Astron.J., 57, 3.

Newton, K. (1980a). Monthly Not. R. Astron. Soc., 190, 689.

Newton, K. (1980b). Monthly Not. R. Astron. Soc., 191, 615.

Seiden, P.E., Elmegreen, D.M. and Elmegreen, B.G. (1987), in preparation.

Seiden, P.E. and Gerola, H. (1982). Fundamentals of Cosmic Physics, 7, 241.

QUESTIONS

LILLER: Were the fake color photographs calibrated - i.e., did you take into account exposure time differences?

ELMEGREEN: Yes, they were calibrated. However, they are fake in the sense that they were normalized so that the composite image would be aesthetically pleasing and scientifically more useful than the standard 3-color photo, which results in a washed-out yellowish image. We purposefully made the blue regions appear bluer, for example, so that young star forming regions would stand out. The grayscales were selected for each galaxy by normalizing the light intensity to be the same in each color between $R=1/3$ and $R=2/3$ times R_{25}.

LIEBOVITCH: Did you ever see any cases where the infrared bar was offset with respect to the blue bar?

ELMEGREEN: No.

(*Photo by* Marjorie Nichols)

Bill Liller with four of his Ph.D. students,
(left to right) Debbie Elmegreen, Phyllis Lugger,
Christine Jones, and Bill Forman.

THE DISCOVERY OF HOT CORONAE AROUND EARLY TYPE GALAXIES

W. Forman and C. Jones
Harvard/Smithsonian Center for Astrophysics, 60 Garden St.,
Cambridge, MA

ABSTRACT. This review describes both the discovery and more recent observations of hot coronae around early-type galaxies. The observations show that early-type galaxies have 0.5-4.5 keV luminosities up to several 10^{42} ergs sec^{-1} (H$_0$ = 50 km sec^{-1} Mpc^{-1}) dominated by thermal emission from 10^{10} M$_\odot$ of hot ($\sim 10^7$K) gas. The cooling time of the gas is sufficiently short that cooling flows as large as a few solar masses per year are common. Finally, the properties of the coronae imply that galactic winds are not now common around early-type galaxies and that the stellar matter of these galaxies is embedded in a massive dark halo extending to ~ 100 kpc.

HISTORICAL REVIEW

The history of the study of gas in early type galaxies began at least as early as 1957. In that year Sandage (1957) estimated that during the course of normal stellar evolution, the constituent stars of a bright elliptical would shed $5 \times 10^9 M_\odot$ of gas. Sandage suggested that much of this gas could be in the form of neutral hydrogen although a portion of it could be ionized and might explain the optical emission lines of forbidden oxygen which were seen in 15% of the ellipticals observed by Humason, Mayall, and Sandage (1956).

In the 30 intervening years, the limits on various forms of gas in early type galaxies became increasingly more stringent and models were developed to explain the absence of gas in elliptical galaxies. For example, Mathews and Baker (1971) discussed the possibility of driving the gas from galaxies in galactic winds powered by supernova explosions. With the thorough review of Faber and Gallagher (1976) it seemed clear that some mechanism must be operating to deplete the gas in early type galaxies. However, Norman and Silk (1979) and Sarazin (1979), whose primary goal was to explain the Butcher-Oemler effect, argued that galaxies with massive halos could maintain gaseous coronae over cosmological times and could suppress the winds which would otherwise result from heating by supernova explosions.

Over the past few years, a clearer picture of bright elliptical galaxies has emerged based on X-ray imaging observations from the Einstein Observatory. These observations have shown the presence of hot gas around bright ellipticals, with gas masses in good agreement with the values suggested in 1957 by Sandage.

EARLY HISTORY OF DISCOVERY OF CORONAE

The first indications of the existence of hot, gaseous coronae around early type galaxies were found in Einstein studies of clusters of galaxies. For example, the earliest imaging observations of the Virgo cluster (Forman et al. 1979) and Abell 1367 (Bechtold et al. 1983) showed the presence of extended emission around early type galaxies. These coronae were found in relatively peculiar environments — rich clusters. Thus, it was not certain whether or not these cluster galaxies owed the presence of their coronae to their environment. Over the past few years, however, evidence has increased indicating the common occurrence of hot coronae around galaxies both in and out of clusters. Coronae were detected around the unusual radio galaxy Centaurus A (Feigelson et al. 1981) and around the S0 galaxies NGC5846 and NGC3607 which are members of small groups (Biermann and Kronberg 1983 and Biermann, Kronberg, and Madore 1982). Still the role of the environment remained uncertain. The report by Nulsen, Stewart, and Fabian (1984) of diffuse emission around a relatively isolated early type galaxy, NGC1395, suggested that the coronae, while they could be influenced by the environment, also could be present around field galaxies. A comparison of samples of early and late type galaxies by Trinchieri and Fabbiano (1985) suggested that the emission mechanisms for these two classes of galaxies were different. Stanger and Schwarz (1984) also concluded, in studies of isolated galaxies, that thermal emission from a diffuse hot gas could be a significant source of X-rays.

To follow up the early observations of coronae in clusters, we initially surveyed 55 early type galaxies with the Einstein Observatory (Forman et al. 1985). We have now expanded that survey to include 180 galaxies. None of these galaxies is the central galaxy in a rich cluster or compact group. We had three basic goals in undertaking these studies. These were:

1. To determine the general X-ray properties of early type galaxies as a class.

2. To measure the gas densities, masses, and temperatures of the coronae.

3. To investigate the underlying gravitational mass in early type galaxies.

To summarize the results of our surveys, we have plotted the X-ray luminosities versus optical luminosities (absolute blue magnitudes) for the 180 early-type galaxies in our enlarged sample (Jones et al. 1988). Figure 1 shows a clear correlation of X-ray with optical luminosity. We also show the results of model calculations for four galaxies with magnitudes -19.5, -20.5, -21.5, and -22.5 from the work of Forman, Jones, and Tucker (1988). Their models follow the evolution of a galaxy and include a stellar component, a dark halo, a hot corona, and a cool interstellar medium. The material lost by evolving stars is sufficient to explain the observed gas masses and X-ray luminosities. The hot gas is heated by supernovae and by the random motions of the stars which are ejecting mass into the corona. The existence of luminous coronae implies that the gas cannot be flowing out of the galaxy in a galactic wind unless we wish to adopt mass loss rates from stars which are greatly in excess of presently accepted rates. Also, simple arguments show that the total gravitational

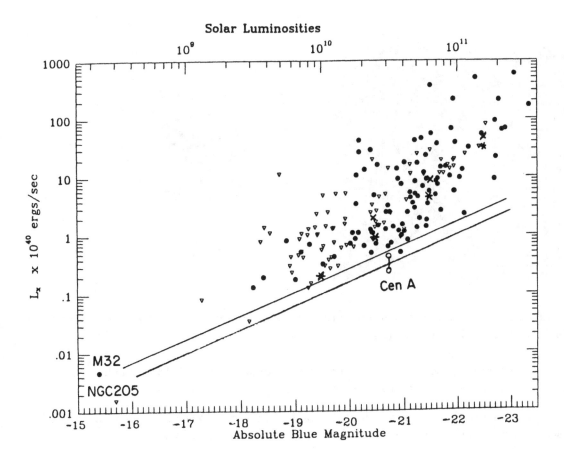

Figure 1: The distribution of X-ray luminosity vs. optical luminosity for ≈180 early-type galaxies. The diffuse emission from Cen A (open circles) is estimated to be no more than 25% to 50% of the total observed emission and is used to limit the possible contribution from galactic sources (lower solid line). The distance to Cen A is taken to be 3 Mpc as derived by Hesser *et al.* (1984). A second larger estimate for the possible contribution from discrete sources is shown as a second solid line (Trinchieri and Fabbiano 1985). Finally, model calculations of galactic evolution (Forman, Jones, and Tucker 1988) are shown as stars and crosses for galaxies of different absolute magnitudes. The symbols represent models with different supernova rates — x=1/3 and star=1/7 of Tammann's (1982) rate — which bracket the newly determined values of van den Bergh et al. (1987). Note that the model of the galaxy of magnitude -19.5, with a supernova rate 1/7 that of Tammann's rate, produces a wind near the present epoch (after 1.2×10^{10} years). The model of a galaxy of magnitude -19.5 with the higher supernova rate does not survive beyond several 10^9 years and is not shown on the figure.

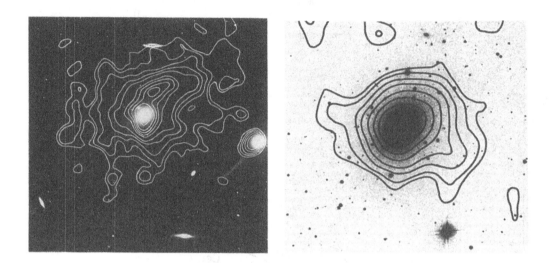

Figure 2: X-ray iso-intensity contours are shown superposed on an optical photograph for M86 (left) and NGC4636 (right).

mass necessary to gravitationally bind the hot coronae is a few $10^{12} M_{\odot}$ which implies mass-to-light ratios of up to 100 (in solar units).

The remainder of this paper will describe the basic observations and the assumptions used to derive the results enumerated above and their implications.

OBSERVATIONS AND BASIC PROPERTIES

The analysis of the X-ray observations of the early type galaxies consisted of four basic elements. First, we generated iso-intensity contour maps (Figure 2). Next, we determined surface brightness profiles of the X-ray emission (Figure 3). These profiles were fitted to a function of the form

$$S(r) = S(o)(1 + (r/a)^2)^{-3\beta + 1/2} \qquad (1)$$

(after convolution with the instrumental response). This form is particularly convenient in deriving an underlying density distribution when the emission arises from an isothermal gas. This analysis was performed for galaxies having sufficient net counts to allow a meaningful determination of the two free parameters a and β. The resulting fits yielded measured values of a ranging from 1.5 to 3 kpc, although several profiles yielded only upper limits (which ranged up to almost 10 kpc). All the observations were consistent with values of a few kpc. The best fit values of β were 0.5 with a range of about 0.1 and again several upper limits ranged up to $\beta \approx 1.0$. All the values of β were consistent with $\beta = 0.5$. It is interesting that this same value is derived for the galaxies at the centers of clusters like M87 in Virgo and NGC 4696 in Centaurus although the gas in rich clusters is best described by a larger β of about 2/3 (Jones and Forman 1984).

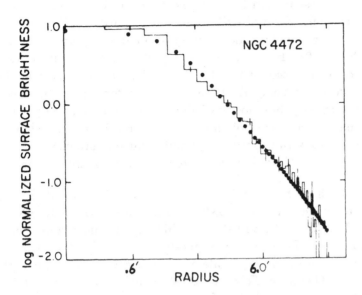

Figure 3: X-ray surface brightness profile of NGC4472 is shown for the energy band 0.5-4.0 keV.

As part of the analysis of the spatial extent, we also tested the ratio of the counts in a small and a large ring around ~100 galaxies. This simple test can provide limited information about the spatial extent. For more than half of the galaxies this ratio indicated that the distribution of the photons was not consistent with a single point source. The remaining galaxies were generally too weak for this test to be useful although the X-ray emission from a few (e.g., M32) is primarily due to a single source. As a final element in the data analysis, we used the IPC pulse height information to perform spectral fits for about 20 galaxies. The measured values of the temperature ranged from 0.5 keV to 4 keV.

The spatial extents and spectral properties provide strong evidence in favor of the thermal mechanism for emission. First, the X-ray spectra themselves are consistent with thermal emission from hot gas of temperatures 0.5-2 keV. Those galaxies with sufficient statistics are inconsistent with power law spectra which would be characteristic of non-thermal emission mechanisms. While some of the galaxies we studied are radio sources, the high resolution imager (HRI) observations place strong limits on a possible point source contribution for four of these. Thus, the evidence for a substantial contribution to the X-ray emission from an active galactic nucleus is in general minimal.

A further argument in favor of a thermal, rather than discrete source, origin for the emission around early type galaxies is based on the steep relation between the X-ray and optical luminosities. For spiral galaxies the X-ray and optical luminosities are linearly proportional. However, for early type galaxies the relation is steeper (e.g. Trinchieri and Fabbiano 1985).

What about the integrated emission from populations of known X-ray sources? One argument is that the optical and X-ray distributions differ spatially. This is most clearly seen for M86 where the x-ray plume does not correspond to any known optical distribution of objects (Figure 2). In addition, the expected X-ray luminosities of the known classes are too low to account for the observed emission. For example, globular clusters can provide at most only about 10^{40} ergs sec^{-1} for a luminous galaxy like M84 or M86 using the average luminosity of globular clusters derived for M31 (which is 3 times higher than the average globular cluster luminosity measured in our own Galaxy). Furthermore, the X-ray spectra of globular clusters are characterized by a much harder spectrum with kT $\approx$5 keV compared to 1 keV for the coronae.

A second possible source of diffuse X-ray emission is from stars. This possibility was discussed and eliminated by Feigelson et al. (1981) in their study of the diffuse X-ray emission in Centaurus A. They noted that M stars would be the primary contributors because of their large numbers. The evidence suggests that Pop II M dwarfs, the relevant population in Cen A and early type galaxies, can account for at most 10% of Cen A's X-ray emission. Therefore, Feigelson et al. concluded that the diffuse emission was most likely produced by radiation from a hot gas. Since Centaurus A is underluminous in diffuse X-ray emission compared to other bright early type galaxies (see Figure 1), it can be used to constrain the contribution of discrete sources to the emission from early type galaxies. For example, if we attribute one-quarter of the diffuse emission from Cen A to discrete sources then the discrete source contribution is 3.5×10^{37} ergs sec^{-1} per $10^9 L_\odot$. For an $M_B = -19$ galaxy this implies less than 2.2×10^{38} ergs sec^{-1} from discrete sources while these galaxies have luminosities of about 10^{39} ergs sec^{-1}. Also, the lack of detected discrete sources in Cen A argues for a small contribution to the diffuse emission. Feigelson *et al.* give an upper bound for point sources of 1.1×10^{38} ergs sec^{-1} (at a distance of 3 Mpc). Thus, sources like those dominating the bulge emission of M31, with luminosities exceeding 10^{39} ergs sec^{-1} (Van Speybroeck *et al.* 1979) cannot be contributing to the extended emission in Cen A. Thus, at the low luminosity end of our dynamic range (M_B=-19), discrete sources could provide no more than about 25% (and possibly considerably less) of the observed emission. This value agrees with the limit derived from the integrated properties of the bulge of our Galaxy (total optical and total X-ray luminosity) and the limits derived from very low luminosity early type galaxies such as NGC205. However, Trinchieri and Fabbiano (1985) have argued that the contribution could be larger based on extrapolating the properties of the bulge of M31 which is more X-ray luminous (per unit optical luminosity) than either the bulge of our Galaxy or Cen A. (A recent detailed analysis of M31 shows a weak, diffuse component which may be due to hot gas.) In conclusion, as shown in Figure 1, the bright galaxies (i.e., $M_B < -20$) are certainly not dominated by emission from discrete sources while at $M_B \approx$-19 the contribution could begin to become important.

What are the properties of the hot gas in the coronae of early type galaxies? Given our surface brightness profiles and the relatively small dispersion in the derived parameters, we have assumed that, in general, the structure of the hot coronae can be described as having core radii of $\approx$2 kpc with $\beta = 0.5$ out to some limiting radius which should be ≈100 kpc (the maximum distance to which we detect the more luminous coronae). With a surface brightness profile and an X-ray luminosity, computed from the observed flux assuming a temperature of 1 keV and distances from Sandage and Tammann(1981), we can derive central gas densities and gas masses. Typical central densities are about 10^{-2} cm^{-3} and the derived masses range from 5×10^8 to $5 \times 10^{10} M_\odot$ depending primarily on the distance to which we were able to trace the X-ray surface brightness profile.

These parameters have interesting implications for the coronae. First, the relatively large observed gas masses imply that the gas must be nearly in hydrostatic equilibrium — it can not be freely flowing out of the galaxy or freely collapsing. If one takes the sweeping time from the galactic wind models of Mathews and Baker (1971), the derived mass loss rates are ≈10 times higher than the predicted values. Thus, if the coronae are not a transient phenomenon, we would have to modify in a significant way the present views of mass loss from evolving stars if the gas were not in hydrostatic equilibrium. As for infall, the gas cannot cool faster than it can radiate away its energy. This allows us to compute a maximum infall velocity of the cooling gas. Outside the core of the gas distribution we find that the infall velocity, V_{infall}, and the sound speed, c_s, are related by

$$(V_{infall}/c_s)^2 = 0.1 \times T_7^{-4.2}/R_{22} \qquad (2)$$

where T_7 is the gas temperature in units of 10^7 K and R_{22} is the radial distance in units of 10^{22} cm. Thus, beyond a few core radii, the gas cannot radiate sufficiently rapidly to achieve a large flow velocity. This constraint applies in the absence of any heat source such as supernova explosions which may provide significant energy to the coronae.

While the outer regions of the coronae are in hydrostatic equilibrium, the cooling times in the central cores can be quite short ($\tau \approx 10^8$ yrs) and, in the absence of heating, accretion flows may occur. Nulsen, Stewart, and Fabian (1984) suggested that cooling flows of $\approx 1 M_\odot$ year^{-1} could be occurring in isolated galaxies. Thomas et al. (1986) analyzed 18 galaxies with hot coronae and found mass accretion rates of 0.02-3.0 $M_\odot$ year^{-1}. Such cooling flows could be important for the initiation of nuclear activity. Also the cooling gas, as it passes through the temperature regime of 10^4 K, could explain the emission lines mentioned by Sandage 30 years ago and studied in detail recently by DeMoulin-Ulrich et al. (1984) and Phillips et al. (1986).

MASSES OF EARLY TYPE GALAXIES

The presence of hot gas in early type galaxies is interesting in its own right. As we have seen, the X-ray emission allows one to determine the observed mass of gas and its distribution. One can estimate the importance of possible cooling flows and galactic winds. For those galaxies in clusters, one can study the interaction of the gas and the intracluster medium. The gas can have important effects on radio lobes emanating from a nuclear region, and the cooling gas could be important in initiating nuclear activity. Also, the coronae, because of their large size, can be important in the interaction of pairs of galaxies especially when one of the galaxies is a gas-rich spiral.

In addition to these phenomena, the gas can tell us something fundamental about early type galaxies — namely their gravitational mass and the shape of the underlying potential which have been difficult to determine for this class of galaxy. The presence of a cool interstellar medium has been used to systematically measure the dynamical masses in spiral systems at large radii with both optical and radio techniques. The absence of such tracers at large radii in early-type galaxies has made the determination of their mass distributions significantly more difficult using optical or radio techniques. However, the constituents of the observed X-ray coronae are ideal test particles to trace the total mass distribution in individual early type galaxies. They obey well-understood laws and comprise a small fraction of the total galaxy mass.

On a large scale, dynamical investigations of elliptical-dominated groups and rich clusters suggest that only a small fraction of the total mass of the system is contained within the optically observable portions of the galaxies and is bound up in stellar systems. The evidence for dark matter is the large mass-to-light ratios of more than 100 derived for the aggregate systems compared to mass-to-light ratios of about 10 estimated for the stellar systems (e.g. Faber and Gallagher 1979). However, the mass distributions of individual early type galaxies remained poorly known.

The use of X-ray emitting gas to trace the gravitational potential was first used by Mathews(1978) and Bahcall and Sarazin (1977) in studying M87. Fabricant et al. (1980) and Fabricant and Gorenstein (1983) used the Einstein imaging observations to accurately determine the underlying mass around M87. Their results showed the power of the X-ray observations to determine the total mass as a function of radius. Well beyond the optically studied region around the galaxy the mass-to-light ratio was found to rise to values in excess of ≈ 100 (in solar units).

The method used to derive the mass distribution in early type galaxies is just that used for M87. This method depends on two simple ideas — the ideal gas law and hydrostatic equilibrium. The hydrostatic equation can be written as

$$\frac{dP}{dr} = -G\rho M(<r)/r^2 \tag{3}$$

where P is the pressure, ρ is the gas density, G is the gravitational constant, and $M(<r)$ is the total mass interior to the radius r. We should emphasize that the mass in question is *not* the mass in gas but all that which contributes to the gravitational potential whether it be stars, collapsed objects, or exotic particles. The ideal gas law allows us to eliminate the pressure in favor of the gas temperature and density and solving for the mass interior to r we find:

$$M(<r) = -\frac{kT}{G\mu m_p}\left(\frac{d\ln\rho}{d\ln r} + \frac{d\ln T}{d\ln r}\right)r. \tag{4}$$

where T is the gas temperature at a radius r, μ is the mean molecular weight and m_p is the mass of a proton. Numerically this becomes:

$$M(<r) = 3 \times 10^{12}(T/10^7 K)\left(-\frac{d\ln\rho}{d\ln r} - \frac{d\ln T}{d\ln r}\right)(r/100kpc)M_\odot. \qquad (5)$$

If one can measure the temperature and density gradients, then one can measure the total mass as a function of radius. This is a remarkably powerful technique and has been applied to a variety of systems which are sufficiently massive to bind gas which radiates at X-ray energies. The scales of these systems range from as small as 10 kpc, to groups and clusters of galaxies, with scales as large as 1 Mpc.

One further comment deals with the spherical symmetry of the systems. Many of the galaxies and clusters are remarkably symmetrical. In the absence of such symmetry one can analyze the observations in sufficiently small angular wedges or by more detailed modelling assuming ellipsoidal mass distributions as has been done for A2256 by Fabricant et al. (1984). These authors found that the more detailed analysis can be used to determine the shape of the underlying potential and shows that the simple analysis gives the mass to within $\approx 20\%$.

Present X-ray observations allow one to readily determine the surface brightness profile which contains information about both the density and temperature profiles. The Einstein imaging proportional counter (IPC) also provides spectral information, although on a cruder scale of a few arc minutes. For sources of sufficient angular size one can thus determine the emission weighted temperature as a function of angular distance projected in the plane of the sky. This combination allows one to both constrain the temperature gradient and to measure the density gradient. This procedure is facilitated by the relative insensitivity of the inferred density distribution to temperature changes. This results partially from the slow change in the emissivity of the gas with temperature i.e., $\Lambda \propto T^{-0.6}$ over the temperature regime appropriate to galaxies. As we also shall argue, the gas outside the central cooling core tends to be isothermal.

The observations we will summarize are those for NGC 4472, a bright Virgo galaxy about 4.5 degrees south of M87. The surface brightness profile from the Einstein imaging detectors, both HRI and IPC, is well determined. We measured the temperature of the coronae in three rings (with outer radii of 3, 5, and 10 arc minutes). We then assumed that the density and temperature distributions in the outer regions could be approximated by power laws. Thus, the exponent of the power laws would give us the logarithmic derivatives — either $\frac{d\ln\rho}{d\ln r}$ or $\frac{d\ln T}{d\ln r}$ — that we require to determine the gravitating mass. For each set of exponents we convolved the density and temperature distribution to determine the emission weighted temperature projected on the sky in the three annuli and the radial surface brightness distribution. Comparison of the predicted distributions with those observed yielded a range of acceptable distributions for the gas temperature and gas density.

In the context of our modeling, the equation for the mass interior to a radius r can be written as:

$$M(< r) = 3 \times 10^{12} \, (\gamma + \alpha) \, T_7 \, r_{100} \, M_\odot \qquad (6)$$

where $-\gamma$ and $-\alpha$ are the logarithmic derivatives of the density and temperature and T_7 is the temperature at the radius r_{100} at which the mass is determined. The smallest acceptable value of T_7 in the outer regions of NGC 4472 is ≈ 1 keV. With this temperature and the most negative allowed value of α (increasing temperature with radius) to cancel γ, we find a mass exceeding $2 \times 10^{12} M_\odot$ at 60 kpc.

Two other galaxies, NGC 499 and NGC 507, have been analyzed in the same manner as NGC 4472 by Jones, Sullivan, and Bothun (1988) and they also find large total masses with no evidence for a steep temperature gradient on scales up to 150 kpc.

Norman and Silk (1979) argued that coronae, like those now detected in X-rays, could be maintained over cosmological times. They also argued that thermal conduction would serve to reduce any temperature gradient since, for sufficiently low density, the cooling timescale exceeds the conduction timescale. The parameters we have determined for the coronae can be applied to their analysis. We require $\tau_{cond} \leq \tau_{cool}$. This inequality holds for radii exceeding R_{crit} :

$$R_{crit}/a = 0.3 \, R_{100}^{2/3} \, T_7^{-1.37} \qquad (7)$$

where R_{100} is the outer radius of the corona, T_7 is the gas temperature, and a is the gas core radius. Thus, beyond a few core radii ($a \approx 2$ kpc), the cooling timescale exceeds the conduction timescale of the gas and we should expect a nearly isothermal atmosphere. Another strong argument for an isothermal corona is that the model calculations which follow the evolution of the coronae over cosmological times find that the coronae are nearly isothermal (e.g. Lowenstein and Mathews 1987).

If we accept the isothermal hypothesis as a general property of the coronae, then we can compute mass-to-light ratios for all those coronae which are resolved. We found mass-to-light ratios ranging from 10 to 90 in solar units. We emphasize that since the derived gravitational mass is proportional to the radius to which the coronae are observed, the actual mass-to-light ratios could be even larger than those we compute. The X-ray measured parameters — the logarithmic derivatives of density and temperature and the value of the temperature - are independent of any assumed distance. Thus, the derived gravitational mass depends linearly on the assumed distance. However, the optical luminosity depends on the square of the distance. Therefore, the mass-to-light ratio varies as (Distance)$^{-1}$. Smaller distances (i.e., a larger Hubble constant) would increase the derived value of M/L.

Fabian et al. (1986) developed a more general method to determine the underlying mass which makes no assumption about the temperature gradient. The method requires hydrostatic equilibrium, stability against convection, and a negative pressure gradient. We have justified hydrostatic equilibrium above and note that the gas must be convectively stable or else convection would return the gas to a stable configuration very rapidly (in a few sound crossing times). The final assumption is justified based on the observed surface brightness profiles. Fabian et al. found that the average total mass (for a sample of 14 early-type galaxies) was $\sim 5 \times 10^{12}$ $M_\odot$. The average mass-to-light ratio for the sample was $\gtrsim 74$. The individual masses ranged from $2 - 7 \times 10^{12}$ $M_\odot$. The masses were roughly twice that found from the isothermal analysis. Thus, the conclusion that individual early-type galaxies are surrounded by dark halos is secure.

COMPARISON OF ELLIPTICAL GALAXIES WITH SPIRALS, GROUPS, AND CLUSTERS

Our first comparison is between the gas mass we observe in ellipticals (per unit luminosity) and that found in rich clusters where the constituent galaxies are primarily early type systems. The relatively small amount of gas we see around individual early type galaxies is consistent with the predictions of mass loss from present epoch stellar systems integrated over $\approx 10^{10}$ years. However, at early epochs much more mass was lost. In fact, the total predicted mass loss averaged over time is 50 times higher than the present epoch value. The bulk of this gaseous material would have been generated at early epochs. The roughly solar abundance of iron in the intracluster medium (ICM) suggests that a substantial fraction of the ICM gas was processed through the constituent galaxies. David et al. (1988) and Jones and Forman, this volume, have shown that in rich clusters about 80% of the ICM could be primordial. Taking $M_{virial}/L_B \approx 300$ for rich clusters with a gas mass of 10% of the virial mass implies $M_{gas}/L_B \approx 30$. To explain the iron abundance in clusters, if 20% of the gas came from the constituent galaxies, then an amount of gas equivalent to $M_{gas}/L_B \approx 6$ also must have been produced by early-type galaxies outside of rich clusters if early type galaxies in and out of rich clusters are the same. For our galaxy sample, the largest value of $M_{gas}/L_B \approx 0.4$ and the average value is 0.16. Thus, much of the gas produced by galaxies must have been ejected.

We can estimate the contribution of gas from early type galaxies to the intergalactic medium (IGM). Davis et al. (1978) have computed the luminosity density of galaxies. If in the field, 20% of galaxies are early types, we find a density of ejected gas of 3×10^{-9} cm^{-3}. Although this density is uncertain, mainly due to the poorly known iron enrichment factor in the ICM, early-type galaxies may be important contributors to the IGM. In one sense, the coronae have solved the problem of where the mass lost by present epoch stars goes. At the same time, the comparison of gas content of rich clusters and early-type field galaxies suggests there may still be a large mass of gas which was ejected into the IGM.

Our second comparison is between elliptical and spiral galaxies. Spirals do not have hot luminous coronae as do early type galaxies in spite of their higher supernova rate per unit luminosity. As discussed earlier, surveys of spiral galaxies have shown that the X-ray luminosity and optical (blue) luminosity are linearly related (e.g. Long and Van Speybroeck 1983 and Trinchieri and Fabbiano 1985), suggesting that the X-ray emission from spirals is predominantly from the integrated contribution of many individual sources. Detailed studies of nearby galaxies (M31, Magellanic clouds : see

Long and Van Speybroeck 1983 for a review and references) have been able to resolve the many sources that produce the integrated emission of spirals.

Searches for hot coronae around four individual spirals have resulted in X-ray upper limits. (Bregman and Glassgold (1982) observed the edge-on systems NGC 4233 and NGC 3628; McCammon and Sanders (1984) analyzed M101; Van Speybroeck has searched for diffuse emission from M31, although a recent analysis shows diffuse emission in the bulge.) The X-ray luminosity limits range from 10^{39} ergs sec^{-1} for M31 to 10^{41} ergs sec^{-1} for M101 and depend on the assumed temperature of the corona. For the same optical luminosity, the spirals have either less X-ray luminous or cooler coronae than their early type counterparts. (At sufficiently low temperatures the Einstein detectors would be insensitive to coronal emission such as that observed in the UV.)

In the plot of X-ray luminosity against blue magnitude (Figure 1), the Sa galaxies tend to fall below the SO and E galaxies. That is, at the same optical luminosity, the Sa galaxies are less X-ray luminous than comparable E and SO galaxies. For both late-type spiral galaxies and Sa galaxies, we have suggested that these galaxies with cool interstellar mediums cannot power extensive hot coronae. Instead, the supernova energy is radiated by the cool medium in the IR and optical where the supernova luminosity is a small fraction of the total emitted radiation. Thus, as one progresses to later type galaxies, the energy available to power a corona decreases.

Another important aspect of the comparison of spirals and early types relates to their dark matter halos. For spirals, the constancy of their rotation curves at large radii implies a dark halo with the gravitating mass proportional to the radius. Since $v^2/r = GM/r^2$ and for $v(r) = constant$ (as is observed), then $M(< r) \propto r$. The gravitational mass in early type galaxies also may have the same radial dependence. At large r, the surface brightness is well described by a power law. If the coronae are isothermal, or nearly so, then the logarithmic derivatives are constant with radius and $M(< r) \propto r$.

This type of radial dependence is predicted from the violent relaxation of dissipationless matter. Thus, a halo of non-baryonic matter (such as those predicted by current Grand Unified Theories in conjunction with the Inflationary scenario) could naturally give rise to the $M(< r) \propto r$ halos of early and late type galaxies (Primack 1984).

Our final comparison relates to the ratio of total mass to luminous mass. Blumenthal et al. (1984) pointed out that as one examines progressively larger systems, a significant portion of the known baryonic mass is not found in stellar systems which contribute to the commonly measured luminosity. For example, in rich clusters of galaxies the mass in stars (in galaxies) is as little as 25% of the mass of gas which fills the intracluster medium and is seen in X-rays. Thus, these authors argue the most revealing quantity is not M_{virial}/L but M_{virial}/M_{lum} where M_{lum} includes luminous matter detected in any wavelength region. This quantity is the ratio of the total mass to that which must be baryonic in form. They found that M_{virial}/M_{lum} was roughly constant, within the uncertainties, for a variety of systems, including dwarf spheroidals (dynamical mass), the Milky Way, spiral groups, small elliptical groups, and rich cluster cores. The values of M_{virial}/M_{lum} are $\approx 10 - 15$ for these systems. One data point missing from the Blumenthal et al. analysis is that for individual early-type galaxies. The X-ray observations now permit us to begin to rectify this omission. Our small sample of elliptical galaxies gives values of M_{virial}/M_{lum} up to ≈ 15 (if we use a mass-to-light ratio of the stellar matter of 6). The luminous matter is almost totally that of the stellar material since the presently observed gas mass is negligible by comparison. As already mentioned, early type galaxies may have expelled large amounts of gas at early epochs in their evolution based on the amount of gas seen in rich clusters. If early-type field galaxies did produce as much gas per unit luminosity as is seen today in rich clusters, then M_{virial}/M_{lum} for the early type field galaxies could be as low as 7. This estimate for M_{virial}/M_{lum} of 7-15 is in excellent agreement with those determined by Blumenthal et al. for other types of systems.

CONCLUSIONS

We have reviewed the X-ray observations of early type galaxies made with the Einstein Observatory and have seen how useful the imaging observations can be for mapping the total mass distribution around these galaxies. The major results found from the study of hot coronae include:

1. Optically luminous early type galaxies are surrounded by envelopes of hot gas ($T \approx 10^7 K$).

2. The gas comprising the coronae can be understood as the material accumulated by the galaxy over its lifetime from the evolution of its stellar component.

3. The gas is only a small fraction of the total galaxy mass and serves as a probe of the mass distribution in the outer regions of early type galaxies.

4. Mass-to-light ratios of individual galaxies approach those of groups and clusters. This suggests that the dark matter in these larger systems may once have been associated with the individual galaxies but was liberated when the larger systems virialized.

5. M_{virial}/M_{lum} for early type galaxies is comparable to that of all other systems — spirals, groups, clusters.

6. Spirals, unlike ellipticals, do not contain luminous hot coronae.

7. Comparison of M_{gas}/M_{lum} in clusters and individual galaxies suggests that at early epochs, early type galaxies may have liberated sufficient material to produce an IGM of density 3×10^{-9} cm^{-3}.

The same analyses described here for early type galaxies also have been applied to groups of galaxies by Kriss et al. (1982), to central galaxies by Fabricant and Gorenstein(1983) and Matilsky et al. (1985), and to clusters (Hughes et al. 1988).

The exploration and exploitation of the implications of the thermal emission around both individual and ensembles of early-type galaxies has only just begun. The most serious hindrance has been detailed temperature information. For galaxies, the Einstein observatory lacked a detector of high spatial and spectral resolution. For nearby clusters, there was adequate spatial resolution, but the limited band width of the telescope did not permit accurate temperature determinations above a few keV. Future observatories will remedy these deficiencies and allow us to attack with confidence and considerable accuracy the problem of the mass distributions in systems containing early type galaxies where hot gas can be detected.

Personal Note

This paper is an updated version of that presented at the 166th AAS Meeting when the first Rossi Prize by the High Energy Astrophysics Division was awarded "for pioneering work in the study of X-ray emission from early type galaxies." It is appropriate that this paper be given on this occasion in honor of Bill Liller. It was through his influence as a thesis advisor, to both of us, and as a colleague that we learned to *look critically at the data.* The observations of the early type galaxies were predominantly taken, by others, several years before they became available to us through the Einstein Data Bank. By looking at the images it was clear that these galaxies were not simple point sources (or groups of a few point sources). With this realization the remainder of the project flowed naturally.

We wish to take this opportunity to thank Bill for showing us, by his example, that the mundane task of looking at the data, without preconceived notions, is perhaps the most important aspect of observational analysis.

We thank K. Modestino for preparing this manuscript. This work was supported through the Smithsonian Scholarly Studies Program and NASA Contract NAS8-30751.

REFERENCES

Bahcall, J. and Sarazin, C. 1977, Ap.J., 213, 699.

Bechtold, J., Forman, W., Giacconi, R., Jones, C., Schwarz, J., Tucker, W., and Van Speybroeck, L. 1983, Ap.J., 265, 26.

Biermann, P., Kronberg, P.P., and Madore, B.F. 1982, Ap.J.(Letters), 256, L37.

Biermann, P. and Kronberg, P. 1983, Ap.J.(Letters), 268, L69.

Blumenthal, G., Faber, S., Primack, J., and Rees, M. 1984, Nature, 311, 517.

Bregman, J. and Glassgold, A. E., 1982, Ap. J., 564, 138.

Canizares, C., Fabbiano, G., and Trinchieri, G. 1987, Ap.J., 312, 503.

David, L., Forman, W., Arnaud, K., and Jones, C. 1988, preprint.

Davis, M., Geller, M., and Huchra, J. 1978, Ap.J., 221, 1.

DeMoulin-Ulrich, M.H., Butcher, H., and Boksenberg, A. 1984, Ap. J., 285, 527.

Faber, S. and Gallagher, J. 1976, Ap. J. 204, 365.

Faber, S. and Gallagher, J. 1979, Ann. Reviews of Astron. and Astrophys., 17,

135.

Fabian, A.C., Thomas, P.A., Fall, S. M., and White, R. E. 1986, MNRAS, 221, 1049.

Fabricant, D., Gorenstein, P., and Lecar, M. 1980, Ap.J., 241, 552.

Fabricant, D. and Gorenstein, P. 1983, Ap. J., 267, 535

Fabricant, D., Rybicki, G., and Gorenstein, P. 1984, Ap.J., 286, 186.

Feigelson, E.D., Schreier, E.J., Delvaille, J.P., Giacconi, R., Grindlay, J.E., and Lightman, A.P. 1981, Ap.J., 251, 31.

Forman, W., Schwarz, J., Jones, C., Liller, W., and Fabian, A. 1979, Ap.J.(Letters), 234, L27.

Forman, W., Jones, C., and Tucker, W. 1985, Ap. J., 293, 102.

Forman, W., Jones, C., and Tucker, W. 1988, preprint.

Hesser, J., Harris, H., van den Bergh, S., and Harris, G. 1984, Ap. J., 276, 491.

Hughes, J., Yamashita, K., Okumura, Y., Tsunemi, H., and Matsuoka, M. 1988, Ap. J., 327, 615.

Humason, M.L., Mayall, N.U., and Sandage, A.R. 1956, A.J., 61, 97.

Jones, C., Forman, W., and Tucker,W. 1988, in preparation.

Jones, C. and Forman, W. 1984, Ap. J. 276, 38.

Jones, C., Sullivan, W., and Bothun, G. 1988, preprint.

Kriss, G., Cioffi, D., and Canizares, C. 1982, Ap. J., 272, 439.

Long, K. and Van Speybroeck, L. 1983, in Accretion Driven Stellar X-ray Sources, eds. W. Lewin and E. van den Heuvel (Cambridge: Cambridge University Press), 117.

Lowenstein, M. and Mathews, W. 1987, Ap.J., 319, 614.

Mathews, W. 1978, Ap.J., 219, 413.

Mathews, W. and Baker, J. 1971, Ap. J., 170, 241.

Matilsky,T., Jones, C. and Forman, W. 1985, Ap. J., 291, 621.

McCammon, D. and Sanders, W. T., 1984, Ap. J., 287, 167.

Norman, C. and Silk, J.1979, Ap. J.(Letters), 233, L1.

Nulsen, P.E.J., Stewart, G., and Fabian, A.C. 1984, MNRAS, 208, 185.

Phillips, M. Jenkins, C., Dopita, M., Sadler, E., and Binette, L. 1986, AJ, 91, 1062.

Primack, J. R. 1984, SLAC-PUB-3387.

Sandage, A. 1957, Ap. J., 125, 422.

Sandage, A. and Tammann, G. 1981, Revised Shapley-Ames Catalog (Washington:Carnegie Institution)

Sarazin, C. 1979, Astrophysical Letters, 20, 93.

Stanger, V. and Schwarz, J. 1984, preprint.

Tammann, G. 1982, in Supernovae: A Survey of Current Research (eds. Rees and Stoneham)

Thomas, P., Fabian, A., Arnaud, K., Forman, W., and Jones, C., 1986, MNRAS, 222, 655.

Trinchieri, G. and Fabbiano, G. 1985, Ap.J., 296, 447.

van den Bergh, S., Mc Clure, R., and Evans, R. 1987, Ap. J. 323, 44.
Van Speybroeck, L. et al. 1979, Ap.J., 234, L35.

Christine Jones (right) conversing with Lanie Dickel.

THE MORPHOLOGY OF CLUSTERS OF GALAXIES, THE FORMATION EFFICIENCY OF GALAXIES, AND THE ORIGIN OF THE INTRACLUSTER MEDIUM

C. Jones and W. Forman
Harvard/Smithsonian Center for Astrophysics, 60 Garden St.,
Cambridge, MA

ABSTRACT. The luminous material in clusters of galaxies falls primarily into two forms — the visible galaxies and the X-ray emitting intracluster medium. The hot intracluster medium (ICM) is the major observed baryonic component of clusters with a mass equal to or greater than that of the stellar matter. In this paper we review the structure and morphology of rich clusters as inferred from X-ray observations of the ICM. We also discuss changes in the efficiency of galaxy formation for different clusters and the origin of the intracluster medium.

INTRODUCTION

An optical image of a cluster of galaxies shows an overdense region of galaxies. The richest, densest clusters contain predominantly early type galaxies (ellipticals and SO's) while in less dense clusters, up to half the galaxies are spirals. Observations of the velocity dispersion for rich clusters result in mass-to-light ratios of ~ 250 (in V or 325 in B). Thus with $M/L \sim 8$ for the stellar portion of individual galaxies, only about 3% of the total cluster mass is contained within the visible galaxies.

X-ray observations provide a different view from that obtained at visible or radio wavelengths. Although emission from individual galaxies in the cluster is sometimes seen, the primary source of X-ray emission is thermal bremsstrahlung from a hot, intracluster medium. The X-ray luminosities of clusters range from 10^{42} to 10^{45} ergs sec^{-1} ($H_0 = 50$ km sec^{-1} Mpc^{-1}) with gas temperatures of 10^7 to 10^8K, comparable to the equivalent temperatures as measured by the velocity dispersions of the galaxies in the cluster. Central gas densities are typically 10^{-3} cm^{-3} (higher in "cooling flow" clusters). At large radii, the gas density decreases with radius as r^{-2}. In the cores of rich clusters, the mass of gas is $\sim 10\%$ of the total cluster mass. Except at the centers of some clusters, the radiative cooling time of the gas is relatively long ($> 10^{10}$ years) and since the sound crossing time is much shorter, the gas can be treated as hydrostatic. Thus, since the gas is in hydrostatic equilibrium in the cluster gravitational potential, it can be used to determine the total cluster mass distribution. Sarazin (1986) has provided a comprehensive review of the properties of clusters of galaxies.

CLUSTER STRUCTURE AND MORPHOLOGY

Present epoch clusters display a wide variety of dynamical states. Gunn and Gott (1972) noted that, while the dynamical timescale of the Coma cluster (a rich, relaxed system) was less than a Hubble time, most other less dense clusters would have dynamical timescales greater than a Hubble time and, hence, could not be fully relaxed. Thus, one of the necessary descriptions of most cluster classification systems (see Bahcall 1977 and Forman and Jones 1982) is a parameter or property relating to the dynamical timescale (e.g., cluster structure, galaxy population).

The observations of clusters can be understood in a framework of dynamical evolution. The rich clusters with cool gas (few $\times 10^7 K$), low velocity dispersions, and low X-ray luminosities have longer dynamical timescales and are less dynamically evolved. The very X-ray luminous clusters with hot gas ($10^8 K$) and high velocity dispersions have shorter dynamical timescales and are relaxed systems.

However, one property that does not neatly fit into this scenario is the presence of a massive, centrally located galaxy. It had been suggested (e.g., Hausman and Ostriker 1978) that the importance of a central galaxy was directly related to the dynamical stage of a cluster as given, for example, by the Bautz-Morgan classification system. However, the X-ray observations show that there is a class of clusters whose properties are those of dynamically young systems — high spiral fractions, low X-ray temperatures, low velocity dispersions, irregular galaxy distributions — which nevertheless display the presence of a massive, centrally located galaxy. The two best examples of these systems are the Virgo cluster with M87 and Centaurus with NGC4696 (Fabricant and Gorenstein 1983 and Matilsky et al. 1985). Other examples are A262 with NGC703 and A1060 with NGC3311. More recent theoretical studies of cluster dynamics argue that the importance of the central galaxy is determined at early stages of the cluster evolution (Carnevali et al. 1981 and Merritt 1985). Following cluster collapse, little further growth of the central galaxy occurs by galaxy mergers. Thus, the suggestion has arisen that a second parameter — the importance of a central galaxy — be added to the dynamical timescale to generate a two dimensional cluster classification system (Forman and Jones 1982).

Figure 1 graphically illustrates the cluster classification system by comparing the optical image and the X-ray contours. The clusters on the top of the figure (A1367—left and A262—right) have low velocity dispersions, low X-ray temperatures, and high spiral fractions which are indicators of systems with long dynamical timescales. The clusters shown in the bottom of the figure (A2256—left and A85—right) have high velocity dispersions, high X-ray temperatures, and low spiral fractions which are indicators of dynamically more evolved systems (those with shorter dynamical timescales). The clusters on the left have no central dominant galaxy while those at the right have a bright galaxy around which the X-ray emission is clearly centered and concentrated. The general characteristics of these clusters are summarized in Table 1.

In the two dimensional cluster classification (see Table 1 and Figure 1) cooling flows appear to be associated exclusively with those clusters having central dominant galaxies, regardless of the dynamical stage of evolution. Thus, dynamically young and old clusters like A1367 or A2256 — with no central dominant galaxy — show no evidence

Table 1 - Cluster Morphology

	Clusters With No Central Dominant Galaxy (nXD)	Clusters With A Central Dominant Galaxy (XD)
Irregular	Low L_x ($< 10^{44}$ ergs/sec) Cool ICM (few $\times 10^7$K) Low velocity dispersion High spiral fraction ($> 30\%$) Low central galaxy density Low central gas density (no cooling flows) examples: A1367, A1314	Low L_x ($< 10^{44}$ ergs/sec) Cool ICM (few $\times 10^7$K) Low velocity dispersion High spiral fraction ($> 30\%$) Low central galaxy density High central gas density (cooling flows) examples: Virgo, Centaurus
Regular	High L_x($> 10^{44}$ ergs/sec) Hot ICM (10^8K) High velocity dispersion Low spiral fraction ($< 30\%$) High central galaxy density Low central gas density (no cooling flows) examples: Coma, A2256	High L_x($> 10^{44}$ ergs/sec) Hot ICM (10^8K) High velocity dispersion Low spiral fraction ($< 30\%$) High central galaxy density High central gas density (cooling flows) examples: Perseus, A85

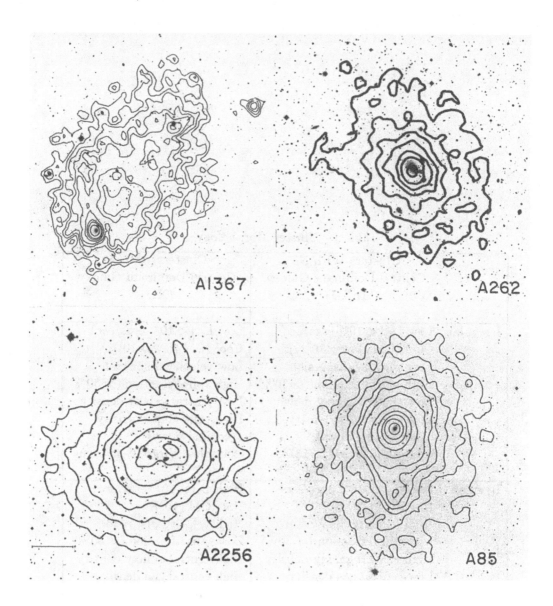

Figure 1: X-ray isointensity contours obtained from Einstein IPC observations are plotted on the optical photographs for the four Abell clusters of galaxies (A1367, A262, A2256, and A85). These illustrate the cluster properties outlined in Table 1.

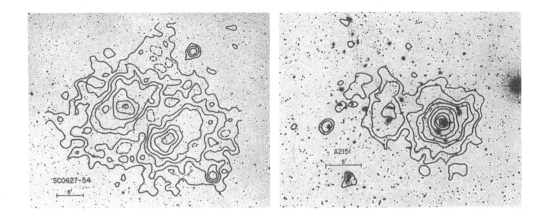

Figure 2: X-ray isointensity contours are shown for two "double" clusters SC0627-54 and A2151 superposed on optical photographs. The X-ray emission which traces the gravitational potential of the clusters shows the subclustering of these systems. Approximately twenty percent of the clusters which could be classified through the X-ray observations have two components.

for cooling gas in their cores while systems like A262 (dynamically young) and A85 (evolved), both show strong evidence of radiative cooling around the central galaxies.

We have now completed an analysis of a sample of about three hundred nearby ($z \lesssim 0.15$) clusters of galaxies observed with Einstein (Jones and Forman 1988). Seventy-five percent of those clusters surveyed were detected in X-rays and most of these were bright enough for a rudimentary classification based on their X-ray morphology. About 70% of these (clusters bright enough for classification) have single peaks in their X-ray surface brightness distributions. Of the remaining 30% with multiple peaks, 2/3 are double and 1/3 are more complex. These multiple-peaked structures are evidence for a still evolving cluster potential (see Cavaliere et al. 1986). Figures 2 and 3 show some examples of the complex systems.

In the Einstein survey about half of the clusters that are single peaked have small core radii ($\sim$ .25 Mpc) and bright galaxies at their centers (Jones and Forman 1984). By comparison about 20% of the Abell sample are classed as cD clusters or Bautz-Morgan I or I-II. For the Abell sample, the X-ray observations show that 40% (twice the number of cD or BM I clusters) are small core radius clusters with bright galaxies at their centers. We have asked if the presence of a central bright galaxy is related to the cluster's core radius by looking at the cumulative distribution for these two groups (see Figure 4). We find that the probability that they are drawn from the same parent population is less than 10^{-3} which suggests a relation between the presence of a bright galaxy and the structure of the cluster.

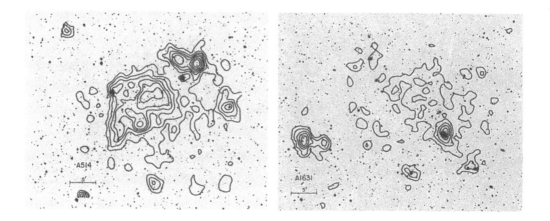

Figure 3: About ten percent of clusters luminous enough to be classified from current X-ray observations show structure which is more complex than the single systems shown in Figure 1 (A262 and A85 in particular) or the "double" clusters illustrated in Figure 2. Shown here are the X-ray isointensity contours for the clusters A514 and A1631 observed with the Einstein IPC superposed on optical photographs.

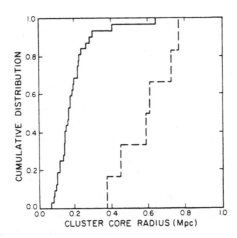

Figure 4: The solid line shows the cumulative distribution of clusters with a central dominant galaxy plotted against the cluster core radius measured from the X-ray observations. The dashed line shows the cumulative distribution for clusters which do not have a central bright galaxy. A K-S test shows that the probability that both distributions were drawn from the same parent population is less than 10^{-3}.

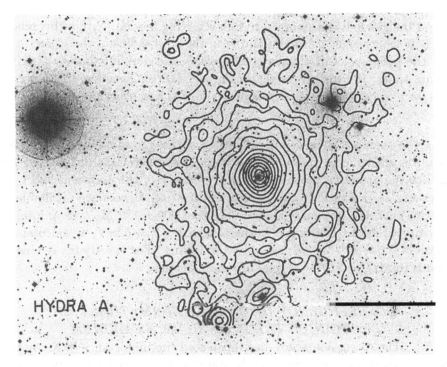

Figure 5: The X-ray isointensity contours obtained from an Einstein IPC observation are plotted on an optical photograph of the cluster around the radio galaxy Hydra A.

The Hydra A cluster can serve to illustrate many of the characteristics of the small core radius clusters (see David et al. 1988a). The Hydra A radio galaxy is the central dominant galaxy in a poor cluster of galaxies not included in Abell's catalog but classed as an Abell richness 0 cluster by Sandage and Hardy (1973). VLA observations (Baum et al. 1988) show a complex radio structure with two knots on either side of a bright core. With an X-ray luminosity of 4×10^{44} ergs/sec, this cluster is typical of luminous rich clusters and brighter than any of the previously studied Morgan groups (Kriss et al. 1983). The X-ray emission as shown in Figure 5 is peaked around the central galaxy.

We use a simple model to describe the X-ray radial surface brightness distribution

$$S(r) = S(o)[1 + (r/a)^2]^{-3\beta+1/2} \tag{1a}$$

where S(o) is the central surface brightness, a is a characteristic scale (core radius), β characterizes the radial fall off of the surface brightness distribution, and r is the radial distance from the cluster center. This formulation can conveniently be inverted to give the underlying gas density distribution (assuming rough isothermality):

$$n(r) = n(o)[1 + (r/a)^2]^{-3\beta/2} \tag{1b}$$

These distributions can be associated with the physical models of a hydrostatic-isothermal gas suggested by Cavaliere and Fusco-Femiano (1976) and Bahcall and Sarazin (1977) where β is the ratio of the specific energy of the galaxies to that of the gas and the galaxies are assumed to follow a King model. Within this model, the characteristic length scale a is identified with the core radius of the galaxy distribution. For the purposes of discussing cooling flows and deriving density distributions and cooling times, the surface brightness model can be considered as a convenient empirical form to derive the gas density distribution and other physical quantities of the gas.

Modelling the surface brightness distribution of the gas in this way for Hydra A, we measure a cluster core radius of 150 kpc. The central gas density is relatively high so that the radiative cooling time at the center of the cluster is short — only 5×10^9 years. Assuming that conduction and heating sources are negligible, David et al. (1988a) determine a mass accretion rate of 600 ± 120 $M_\odot$/yr. Assuming that the gas temperature distribution is isothermal with kT=4.5 keV, we calculate a total virial mass of $3.5 \times 10^{14} M_\odot$ within a 1 Mpc radius. For the Hydra A cluster this corresponds to M/L = 270 — typical of a rich cluster.

THE EFFICIENCY OF GALAXY FORMATION AND
THE ORIGIN OF THE INTRACLUSTER MEDIUM

The discussion above has illustrated the importance of X-ray observations of the intracluster medium to understanding the overall properties of clusters. In addition to providing information on the structure and morphology of clusters, the amounts of accreting material, and the total mass distribution, the study of the intracluster medium is important in its own right. The luminous material in clusters falls primarily into two forms — the visible galaxies and the X-ray emitting intracluster medium. The ICM is a major baryonic component of the cluster being equal or greater in mass than the stellar matter. It is of particular importance to determine the origin of such a large fraction of the known baryonic mass of the cluster. Equally fundamental and related problems to address are the effects of the ICM on the morphology of galaxies and the efficiency of galaxy formation.

To begin to address the questions of the origin of the ICM and the efficiency of galaxy formation in different environments, it is useful to compare the ratio of gas mass to stellar mass in groups of galaxies and rich clusters. It is well known that the mass-to-light ratio increases with the size of the system. However, from poor to rich clusters the fraction of X-ray emitting gas to virial mass remains relatively constant ($\sim 10\%$ within the central five core radii). Therefore, the ratio of gas mass to stellar mass should increase from the poor to rich clusters. As shown in Figure 6, in groups of galaxies the gas mass is approximately equal to the stellar mass, while in very rich clusters, the gas mass exceeds the stellar mass by approximately a factor of six.

The discovery of heavy elements in the ICM (Mitchell et al. 1976 and Serlemitsos et al. 1977) revealed an entirely new aspect to the study of the ICM — one which is crucial in determining its origin. Since heavy elements can be produced only through thermonuclear reactions in stars or by supernovae, the discovery that the intracluster medium was enriched in heavy elements required that material processed through stars be ejected into the ICM. The near solar abundance of the ICM measured by

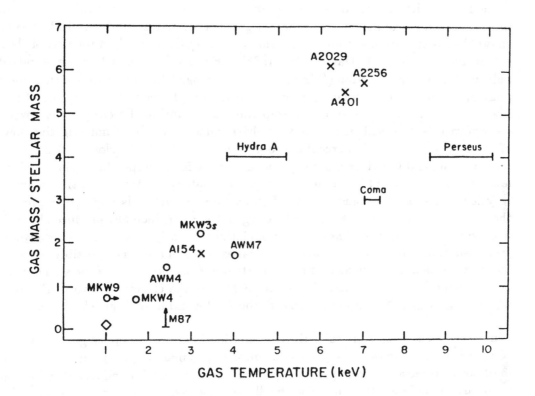

Figure 6: The ratio of the gas mass as measured by the X-ray observations to the stellar mass is plotted against the temperature of the gas. The gas and stellar masses are evaluated within five core radii. The temperature of the gas reflects the gravitational potential of the system. The increasing ratio of gas mass to stellar mass with increasing gas temperature suggests a decreasing efficiency of galaxy formation (conversion of gas to luminous stellar matter) between groups and rich clusters.

early X-ray experiments led to the suggestion that a large fraction of the material in the ICM was ejected from galaxies (e.g. DeYoung 1978).

While the enriched material must come from the galaxies (in the absence of a population III component), more recent studies have suggested that the bulk of the ICM of a rich cluster could not have originated within the galaxies because its mass is several times larger than the mass of the galactic stellar component. Thus, the bulk of the ICM in rich clusters must be "left over" from the formation of the galaxies. In particular, numerical modelling of the hot gaseous coronae around elliptical galaxies shows that over a Hubble time these galaxies can contribute only a fraction of their stellar mass to the ICM (David et al. 1988b). For a galaxy model with a standard Salpeter initial mass function (IMF), stars, on "average," liberate $\approx 40\%$ of their total mass up to the present epoch. However, realistic models show that only 25% of this liberated mass is actually expelled from the galaxy during the early epochs when supernova rates are sufficiently high to drive galactic winds. Changing the slope of the IMF can alter this percentage to 40% for a flatter IMF (slope = 2.0 ; more massive stars and therefore more supernovae) or 1% for a steeper IMF (slope = 3.0 ; fewer massive stars and fewer supernovae). The material remaining in the galaxy is recycled into subsequent generations of stars. Some recycling is required to provide the observed heavy elements (especially oxygen which is principally produced by high mass supernovae) in the older stellar populations of early-type galaxies. However, even assuming that *all* the liberated mass was expelled in a wind or possibly removed by other mechanisms such as ram pressure stripping, at most 40% of the stellar component could have been injected into the ICM. For the clusters shown in Figure 6 where $M_{gas}/M_{stellar} > 2$, less than 50% of the ICM could have originated in galaxies.

While such an analysis constrains the contribution from present epoch galaxies to the ICM, we can use the ratios of the stellar mass to gas mass in groups and clusters to limit both the mass loss from present epoch galaxies to the ICM and any contribution from early, population III stars. Specifically, so long as the IMF's and the population III component of groups and clusters are similar, then we would expect the stellar contribution to the ICM per unit stellar mass to be the same in all groups and clusters. The approximate equality of gas mass and stellar mass in the low X-ray luminosity Morgan groups (MKW4, MKW9, and AWM4) limits the contribution to the ICM by all stars to no more than the present stellar mass. Thus, in the richest clusters where the gas mass is three to six times the stellar mass, only a small fraction of the ICM could have been produced in stars. In the rich clusters, most of the gas in the intracluster medium must be primordial.

The ratio of the gas mass to the stellar mass, $M_{gas}/M_{stellar}$, shown in Figure 6, can be related to the efficiency of star formation. We assume a scenario in which the luminous matter (stars and the ICM) form the bulk of the baryonic material and the remainder of the virial mass is in the form of hot or cold dark matter. Then the efficiency of galaxy formation, the conversion of baryons from gas to stars in galaxies, can be written as

$$\epsilon = M_{stellar}/M_{lum} \tag{2}$$

where $M_{lum} = M_{stellar} + M_{gas}$, or equivalently as

$$\epsilon = (1 + M_{gas}/M_{stellar})^{-1} \qquad (3)$$

(assuming the expelled gas from galaxies is small and can be neglected). Thus by measuring $M_{gas}/M_{stellar}$ we can study the efficiency of star formation in systems ranging from groups to rich clusters. Our analysis shows that the star (and galaxy) formation efficiency ranges from 50% for groups to as little as $\approx$15% for rich clusters. If all the ICM in groups is gas ejected from galaxies, and we use this injection rate for all clusters, then the galaxy formation efficiency would be 100% for groups but still a lower efficiency (as low as $\approx$17% for $M_{gas}/M_{stellar} = 6$ to as high as $\approx$50% for $M_{gas}/M_{stellar} = 3$) for rich clusters. Although the ratio of luminous material (gas+stars) to total gravitating mass remains relatively constant for all clusters (Blumenthal et al. 1984), the efficiency of galaxy formation decreases as one moves to richer systems. In other words although the richest systems obviously produced more galaxies, their efficiency of galaxy formation was lower.

Interpreting the ratio of gas mass to stellar mass as a measure of galaxy formation efficiency requires that clusters be "closed" systems, that is, no material may be added or lost. The gas in the ICM is enriched both during an early phase of massive star supernovae (type II) and continuing through the present with primarily type I supernovae and mass loss from older stars. Since the gravitational potential of poor clusters is sufficient to bind the enriched material ejected in supernova winds driven from the galaxies, none of the material in the ICM should be lost from these systems. Furthermore, based on the computed enrichment rates, extensive amounts of matter could not have been expelled by the galaxies and entirely lost from the rich clusters if their ICM's are to have significant heavy-element abundances. Therefore the change in the ratio of gas mass to stellar mass with cluster richness cannot be explained by a loss of hot intracluster material from the groups and poor clusters. The relative constancy over rich and poor clusters of the fraction of the cluster virial mass made up by luminous materials (stars and gas) also supports the notion of a "closed" system.

In such a "closed" system as long as the galaxies in different clusters have similar initial mass functions so that the amount of enriched material is directly related to the stellar mass, then the decrease in efficiency of galaxy formation between groups and rich clusters implies that the intracluster medium in poor clusters should show a greater abundance of heavy elements than that in rich clusters. Since the conclusion concerning different enrichments of the ICM hinges on clusters and groups having similar IMF's, it is important to note that Morgan, Kayser and White(1975) found the galaxy populations of many of the cD dominated groups to be similar to those of rich Abell clusters of Bautz-Morgan Type I. There is no evidence that the galaxies in rich clusters are significantly different from the galaxies in these dense groups.

Since it is likely that the galaxies in rich and poor clusters formed with similar IMF's, the iron (and other heavy elements) produced during the evolution of the component stars would yield a mass of iron directly proportional to the present epoch stellar light of the galaxies:

$$M_{Fe} = \eta L_{galaxies} \tag{4}$$

where η is determined by the IMF. Let us assume that the fraction of material ejected from the galaxies into the ICM is a constant fraction of the material liberated by the constituent stars. This is supported observationally by the similar colors of Morgan group galaxies and cluster galaxies implying comparable metal abundances. Further support comes from the similarity in the effectiveness of gas removal mechanisms. Internal mechanisms (e.g. supernova driven galactic winds) would be indifferent to the environment and ram pressure stripping (proportional to ρv^2) of the gas from galaxies is comparable in groups and clusters (the gas temperature, a measure of v^2, increases by a factor of 3-4 from groups to rich clusters and the central gas densities are perhaps slightly higher in groups and hence partially offsetting). Thus, in the simplest scenario we can predict a trend of iron abundance with the depth of the potential of the system — increasing potentials, correlated with lower galaxy formation efficiencies, imply decreasing iron abundances.

The present measurements of iron abundances are too inaccurate to verify the above model or test possibilities for the origin of the ICM. Mushotzky (1984) and Henriksen (1985) summarize present results. For rich clusters, Henriksen (1985) reports a possible correlation of decreasing iron abundance with increasing gas temperature, as predicted, but the data are not sufficiently precise to yield quantitative results. Those observations were for only quite luminous clusters ($L_x > 2 \times 10^{44}$ ergs sec^{-1}) while in general we expect the abundances to be highest in the low luminosity clusters. Hughes et al. (1988) have measured a precise iron abundance of 22% of the solar value for the rich Coma cluster. To adequately test for differences in abundances, it is particularly important to obtain comparable measurements for low temperature (high galaxy formation efficiency) systems. By determining accurate values of the heavy-element abundances of the ICM in both poor and rich clusters, one could better investigate the properties of the IMF (e.g. exponent), the efficiency of galaxy formation, and the origin and enrichment of the ICM. A precise determination of the heavy element abundance of the intracluster medium for a sample of clusters ranging from groups to rich clusters has implications for the amount of material in the ICM that must be primordial. In particular, determining a high iron abundance for the ICM in Morgan groups, as suggested by the arguments above, would confirm that the origin of most of the hot gas in rich clusters must be primordial.

The changing ratio of gas mass to stellar mass also will affect the energy (or temperature) of the ICM. By measuring the surface brightness profiles and independently by measuring the ratio of the velocity dispersion to the gas temperature, one can estimate the energy per unit mass of the galaxies compared to that of the gas. From the surface brightness profiles, this value for rich clusters is generally $\sim 2/3$. The values calculated from the measured velocity dispersions and gas temperatures have a wider range (but see Flanagan et al. 1988 who suggest a resolution for the Perseus discrepancy). By comparison to rich clusters, the surface brightness profiles for hot gas around single dominant cluster galaxies such as M87 and the cD groups such as AWM7 yield a value $\sim 1/2$ (Fabricant and Gorenstein 1983). This implies that the groups and individual central galaxies have more energy per unit mass in gas compared to the constituent galaxies than do rich clusters. David et al. (1988b) have estimated the additional heat that would be added to the ICM from supernova driven winds which occur during early epochs when supernova rates are high. They find that the fractional increase in the gas temperature is given by:

$$\frac{T_{gas} - T_{initial}}{T_{initial}} = \frac{M_{ej}}{M_{ej} + M_{initial}} \left\{ \frac{T_{ej}}{T_{initial}} - 1 \right\} \tag{5}$$

where $T_{initial}$ is the initial temperature of the ICM, T_{gas} is the present epoch temperature of the ICM, T_{ej} is the temperature of the ejected gas, M_{ej} is the total mass ejected from galaxies, and $M_{initial}$ is the mass of the ICM which remains after galaxy formation. This expression can be rewritten as:

$$\frac{T_{gas} - T_{initial}}{T_{initial}} = f_{ej} \frac{M_{stellar}}{M_{gas}} \left\{ \frac{T_{ej}}{T_{initial}} - 1 \right\} \tag{6}$$

where f_{ej} is the fraction of the stellar mass ejected from a typical galaxy, $M_{stellar}$ is the stellar mass in galaxies in a cluster, and M_{gas} is the total present cluster gas mass. Calculations show that f_{ej} lies in the range 0.1-0.4 and $T_{ej}/T_{initial}$ ranges from 2-5. Therefore for the groups and poor clusters where the stellar mass is comparable to the gas mass, there may be significant heating of the ICM by the ejected material which may account for the observed difference between the groups and the clusters.

SUMMARY

This paper has highlighted some of the properties of clusters of galaxies in which X-ray observations have been particularly useful. These include:

1. the recognition that clusters of galaxies exhibit a variety of morphological forms which can be understood within a framework of dynamical evolution

2. the discovery of an increase in the ratio of gas mass in the intracluster medium to stellar mass as one compares groups to rich clusters implying a decrease in galaxy formation efficiency from poor to rich systems

3. the conclusion, also based on the large ratio of gas mass to stellar mass in rich systems, that most of the intracluster medium in these clusters is primordial.

Acknowledgements

At this symposium in honor of Bill Liller, it is particularly appropriate to acknowledge the substantial contributions he made toward our astrophysical training. Much of our work both on galaxies and on clusters of galaxies have used X-ray observations which, as part of a data bank, were available to (and often originally observed by) others. Our success in gaining new information from these observations stems from an ability to recognize, at least occasionally, interesting problems and to see (sometimes) in the data new aspects or properties of the source. We owe these skills in large part to Bill Liller.

We thank Karen Modestino for her excellent preparation of this manuscript. This work was supported through the Smithsonian Scholarly Studies Program and NASA Contract NAS8-30751.

REFERENCES

Bahcall, N. 1977, Ann. Rev. Astron. Astrophys., 15, 505.

Bahcall, J. and Sarazin, C. 1977, Ap.J., 213, L99.

Baum, S., Heckman, T., Bridle, A., van Breugel, W., and Miley, G. 1988, preprint.

Blumenthal, G.R., Faber, S.M., Primack, J.R., and Rees, M.J. 1984, Nature, 311, 517.

Carnevali, P. Cavaliere, A., and Santangelo, P. 1981, Ap.J., 249, 449.

Cavaliere, A., and Fusco-Femiano, R. 1976, Astron. Astrophys., 49, 137.

Cavaliere, A., Santangelo, P. Targuini, G., Vittorio, N. 1986, Ap.J., 305, 651.

David, L. Arnaud, K., Forman, W., and Jones, C. 1988a, preprint.

David, L., Forman, W., and Jones, C. 1988b, preprint.

DeYoung, D.S. 1978, Ap.J., 223, 47.

Fabricant, D. and Gorenstein, P. 1983, Ap.J. 267, 535.

Flanagan, J., Hughes, J., Arnaud, K., Forman, W., and Jones, C. 1988, in preparation.

Forman, W. and Jones, C. 1982, Ann. Rev. Astron. Astrophys., 20, 547.

Forman,W., Jones, C., and Tucker, W. 1985, Ap.J., 293, 102.

Gunn, J.E. and Gott, J.R. 1972, Ap.J., 176, 1.

Hausman, M.A. and Ostriker, J.P. 1978, Ap.J., 224, 320.

Henriksen, M. 1985, Ph.D. Thesis (University of Maryland).

Hughes, J.P., Yamashita, K., Okumura, Y., Tsunemi, H., and Matsuoka, M. 1988, Ap.J., 327, 615.

Jones, C. and Forman, W. 1984, Ap.J., 276, 38.

Jones, C. and Forman, W. 1988, in preparation.

Kriss, G., Cioffi, D., and Canizares, C. 1983, Ap.J., 272, 439.

Matilsky, T., Jones, C., and Forman, W. 1985, Ap.J., 291, 621.

Merritt, D. 1985, Ap.J., 289, 18.

Mitchell, R., Culhane, J.L., Davison, P.J., and Ives, J.C. 1976, MNRAS, 176, 29p.

Morgan, W.W., Kayser, S., and White, R.A. 1975, Ap.J., 199, 545.

Mushotzky, R. 1984, Phys. Sci., T7, 157.

Sandage, A. and Hardy, E. 1973, Ap.J., 183, 743.

Sarazin, C. 1986, Reviews of Modern Physics, Vol. 58, 1.

Serlemitsos, P.J., Smith, B.W., Boldt, E.A., Holt, S.S., and Swank, J.H. 1977, Ap.J., 211, L63.

Speaker: Christine Jones

Question from D. Harris:

The previous speaker (W. Forman) indicated that the hot gas in ellipticals probably comes from stellar evolution. For clusters, we expect that most of this gas eventually joins the ICM. Is the <u>quantity</u> of heavy elements (inferred from Fe lines) in the ICM consistent with that produced by stellar evolution in the member galaxies?

By: D. Harris.

Response:

The measured iron abundances for the ICM are consistent with that produced by stars in the member galaxies. However these abundances are poorly known. Only the Coma cluster has a good determination of its iron abundance. As we have emphasized in this paper, determining the iron abundance for a variety of clusters has important implications for galaxy formation and the origin of the intracluster mediums.

(*Photo by* Marjorie Nichols)

Phyllis Lugger conferring with Jim Elliot.

TESTING MODELS FOR THE DYNAMICAL
EVOLUTION OF CLUSTERS OF GALAXIES

Phyllis M. Lugger
Department of Astronomy
Indiana University
Bloomington, IN 47405 USA

Abstract. I review theoretical and observational studies concerning the dynamical evolution of clusters of galaxies. The theoretical studies have investigated the effects of a number of gravitational interaction mechanisms including violent relaxation, two-body relaxation, dynamical friction, galaxy mergers, and tidal stripping. A central issue addressed by many of these studies is the possible production of supergiant cD galaxies by repeated galaxy mergers in dense cluster cores; this process has been termed "galactic cannibalism." I describe observational studies designed to test the predictions of the theoretical models. These studies have investigated several properties of galaxies in clusters, including: (1) the colors and structure of first-ranked galaxies, (2) the galaxy luminosity function, and (3) the dynamics of multiply nucleated first-ranked galaxies. The observational results indicate that the effects of dynamical evolution on clusters are more subtle than originally predicted.

1   INTRODUCTION

In 1975 Bill Liller suggested that I study the association between unidentified high galactic latitude *Uhuru* X-ray sources and clusters of galaxies and thus began my work in the exciting field of the observational study of clusters of galaxies. With Bill's guidance, I carried out this work for my undergraduate thesis project at Harvard (Lugger 1978). I was again fortunate to have Bill advise my Ph.D. dissertation at Harvard on observational evidence of dynamical evolution in galaxy clusters (Lugger 1982, 1984a,b, 1986). For the latter work, Bill was instrumental in arranging for numerous observing runs on the Palomar Observatory 1.2 m Schmidt telescope and the Kitt Peak National Observatory (KPNO) 0.9 and 1.3 m telescopes. He cheerfully accompanied me on many Palomar observing runs, where I greatly benefited from his vast expertise in photographic technique.

1.1   *Properties of Clusters of Galaxies*

Clusters of galaxies are systems with radii of several Mpc [1] containing tens to thousands of galaxies. These aggregates represent a fundamental level

---

[1] A parsec (pc) is equal to 3×10^{13} km.

of structure in the spatial distribution of galaxies in the universe. Clusters are observed to exhibit a wide range of morphological types. They vary in their gross structural properties from loose *irregular* associations to centrally concentrated, symmetric *regular* systems. There is a correlated variation in galaxy content with irregular clusters containing a preponderance of spirals while regular clusters are mostly populated with S0 [2] and elliptical galaxies.

The morphological properties of clusters have been described by various classification schemes. A commonly used system is that due to Bautz and Morgan (1970; BM), which is based on the luminosity contrast between the brightest galaxy (the "first-ranked galaxy") and other bright galaxies in the cluster. Clusters with a "cD" galaxy — a dominant, supergiant elliptical that has an extended low surface brightness halo (Mathews, Morgan, and Schmidt 1964) — are designated BM I. Clusters in which all of the bright galaxies are similar in luminosity are designated BM III. There are three intermediate classes between these extremes. BM I type clusters are typically regular while BM III type clusters are typically irregular. It has been suggested that the cluster morphological sequence from irregular to regular structure represents a sequence of advancing phases of dynamical evolution (see discussion of cluster virilization in §2.2).

The structure and dynamics of first-ranked galaxies are key data in the study of the dynamical evolution of clusters of galaxies. One striking observation is that the first-ranked galaxy in as many as 50% of all clusters has multiple nuclei, one or more smaller galaxies within $10h^{-1}$kpc projected distance [3] of the nucleus of the dominant galaxy (Hoessel and Schneider 1985). The fraction of first-ranked galaxies with multiple nuclei is not strongly dependent on cluster morphological type. Another important result is that cD galaxies typically move at low velocity relative to the cluster mean, indicating that they are nearly at rest with respect to the center of the cluster potential well.

2   THEORETICAL BACKGROUND

In the following sections, the basic physical interaction mechanisms that drive the dynamical evolution of clusters of galaxies are discussed and the results of simulations of cluster evolution are reviewed. The reader is also referred to the reviews of the physical theory of dynamical evolution of galaxy clusters by Sarazin (1986, 1988) and Merritt (1988).

2.1   Interaction Mechanisms

In the simplest approximation, galaxies in a cluster may be regarded as point masses that undergo two-body gravitational scattering, much like stars in a galaxy. In addition, coupling between translational and internal degrees of freedom

---

[2] An S0 galaxy has a disk as does a spiral, but shows no evidence of spiral structure and contains very little gas and dust, as is the case for an elliptical.

[3] The parameter h is the Hubble constant H_0 measured in units of 100 km s^{-1} Mpc^{-1}. A value of $h = 0.5$ is adopted throughout this paper, i.e. $H_0 = 50$ km s^{-1} Mpc^{-1}.

is also expected for cluster galaxies, since galaxy sizes are appreciable relative to the intergalaxy separation. Such coupling typically converts the orbital energy of galaxies into internal energy, thus "heating" the stellar velocity distribution within galaxies. Two possible outcomes of this energy transfer are galaxy mergers and tidal stripping of mass from the outer parts of galaxies. Following are descriptions of physical interaction mechanisms thought to be relevant to galaxies in clusters: violent relaxation, two-body relaxation, dynamical friction in the intracluster background, galaxy merger, and tidal stripping.

Violent Relaxation and Virialization. The classic picture of the early dynamical evolution of a galaxy cluster is one of gravitational collapse (see e.g. Peebles 1970). Due to gravitational attraction, a region of overdensity condenses out of the general Hubble expansion of the universe and collapses to achieve a state of "virial equilibrium." This state is characterized by a spatial structure that evolves only on a time scale much longer than the collapse time. In virial equilibrium, the motions of galaxies within the cluster provide the "pressure" that balances the inward gravitational force. On the assumption that galaxies form before the collapse has proceeded to completion, there is a phase of violent relaxation during which galaxy velocities are randomized in the strongly fluctuating gravitational potential of the collapsing system. The result of this violent relaxation process is an approximately equal distribution of *specific* kinetic energy ($\frac{1}{2}v^2$ where v is the galaxy velocity), over galaxies of all masses. Since the effect of violent relaxation is independent of galaxy mass, it does not result in a spatial segregation of galaxies by mass. As discussed in §2.2, the picture of a global collapse with a roughly spherical structure is an oversimplification. Detailed simulations indicate the formation and agglomeration of substructures within proto-clusters.

Two-body Relaxation and Dynamical Friction. Two-body relaxation is the process by which galaxy-galaxy gravitational interactions tend to produce energy equipartition, i.e. the approximately equal distribution of random translational kinetic energy ($\frac{1}{2}mv^2$, where m is the galaxy mass) over galaxies of all masses. The predicted long-term result is for the density at the center of a cluster to increase in time, with the most massive galaxies settling there. (In energy equipartition, more massive galaxies have lower typical velocities and thus lower scale heights in the cluster potential.) However, the time required for energy equipartition to be established in a *typical* location in a galaxy cluster is greater than the age of the universe, unless all of the mass of the cluster resides in galaxies (White 1976b). In fact, little mass segregation is observed and thus galaxies appear to account for no more than about 10% of the mass of clusters (Sarazin 1986). Two-body relaxation may, however, be important for the most massive galaxies in the central regions of clusters.

Observational studies of the dynamics of individual galaxies and of clusters of galaxies indicate an increase of mass-to-light ratio (M/L) as the system size increases. Typical estimates of M/L values are about 5 – 10 for the *visible* regions

of individual galaxies, 30 – 100 for individual galaxies measured to large radii
(∼ several × 10 kpc), and about 300 for rich clusters, where a solar type star has
$(M/L)_\odot = 1$ (see articles by Jones and Forman, and Forman and Jones in this
volume). Thus, much of the mass of a rich cluster is in the form of rather low
luminosity material. Hot, X-ray emitting intracluster gas only accounts for about
10% of the cluster mass, comparable to the amount in the visible parts of galaxies.
Suggested forms for the unseen dark matter include "WIMPs" (weakly interacting
massive [subatomic] particles), substellar objects (Jupiter mass objects and brown
dwarfs), and degenerate stellar remnants (white dwarfs, neutron stars, and black
holes). While some of this underluminous material may be associated with the
halos of individual galaxies, it is likely that most of it forms a smoothly distributed
"background" in the potential well of the galaxy cluster (Merritt 1988). This
background may account for about 80% of the mass of the cluster.

Due to the tendency towards energy equipartition, galaxies are, on average, slowed
as a result of gravitational interactions with lower mass background objects. This
effect can be understood in terms of the gravitational wake produced by a "test"
galaxy as it traverses a background of less massive galaxies or stars (see discussion
in Binney and Tremaine 1988). The wake results from the deflection of the orbits of
the background masses toward the trajectory of the test galaxy. The resulting mass
overdensity behind the test galaxy produces a retarding force. A more massive
test galaxy produces a greater overdensity and thus a greater retarding force. The
dynamical friction force also increases strongly with decreasing relative velocity
between the test galaxy and the background masses, since the interaction time
varies inversely with velocity.

Dynamical friction will cause galaxies to spiral in toward the center of the cluster
(Lecar 1975, Ostriker and Tremaine 1975, White 1976a). The rate of orbit decay
due to dynamical friction is proportional to the product of the mass of the galaxy
experiencing the friction and the mean mass density of the background. Thus,
dynamical friction should tend to concentrate the most massive galaxies near the
center of a cluster. Estimates of the degree to which this may have occurred in a
cluster also place upper limits on the fraction of the cluster mass that is in the form
of galaxies (White 1976a).

Mergers. Galaxies passing through the halo of a massive galaxy will experience
dynamical friction in the background of halo material; this process, termed
"galactic cannibalism," can result in the capture of "victim" galaxies by a massive
"cannibal" galaxy (Ostriker and Tremaine 1975). More generally, a close encounter
between any pair of galaxies can result in merger provided that the collision
velocity is not much larger than the internal stellar velocity dispersion in the
galaxies and that the distance at closest approach is less than the sum of the half-
mass radii of the galaxies (Tremaine 1981). Thus mergers tend to occur between
pairs of galaxies that are initially nearly gravitationally bound. In this case, enough
of the relative orbital energy is transferred into internal energy, during a close
encounter, to bind the galaxies. The time scale for the bound pair of galaxies to

merge into a single system is of order the internal stellar orbital time scale, a few times 10^8 years. Since the galaxy velocity dispersion in a rich cluster exceeds the stellar velocity dispersion within a galaxy by a factor of about five, most close galaxy-galaxy encounters do not lead to mergers.

Tidal Interactions. Galaxy encounters that do not lead to mergers tend to increase the internal energy of the galaxies, as a result of the "kicks" experienced by the stars in each galaxy due to the transient tidal force of the other galaxy. (Spitzer [1958] first considered this problem in a different context.) This energy input can cause a loss of stars from the outer parts of the galaxies; this process is termed tidal stripping. An important goal of the theoretical work has been to predict the rate at which tidal stripping reduces galaxy masses. Early treatments of tidal interactions between galaxies were carried out in the impulse approximation, in which the stars are regarded as motionless during the encounter (e.g. Gallagher and Ostriker 1972). This approximation is reasonable when the encounter velocity substantially exceeds the stellar velocity dispersion within the galaxies. More recent work has relaxed the impulse approximation, tracking the motions of individual stars during the course of the encounter (Richstone 1975, Dekel, Lecar, and Shaham 1980, Aguilar and White 1985). This refinement has allowed the treatment of a wide range of impact parameter, from head-on collisions to distant encounters, and a wide range of collision velocity, from low-speed nearly parabolic orbits to high-speed hyperbolic orbits. Close, low-speed encounters produce the strongest perturbations of galaxy structure, due to the high strength of the tidal forces and the long duration over which they operate.

Material stripped from galaxies is expected to join the general background distribution in the cluster (Richstone 1976). Tidal stripping and mergers can be regarded as competing mechanisms for the evolution of *individual* galaxies, in that stripping reduces galaxy masses while merging selectively increases the masses of some galaxies. However, a galaxy that is located at the center of the cluster potential well can grow both by mergers and by accretion of material that has been stripped from other galaxies. These mechanisms have been proposed for the production of cD galaxies.

2.2   *Evolutionary Simulations and Results*

Over the past two decades, computer simulations have been used to investigate the dynamical evolution of galaxy clusters and the effect of environment on galaxies in clusters (e.g. Peebles 1970; White 1976b; Hausman and Ostriker 1978; Richstone 1976; Richstone and Malumuth 1983; Malumuth and Richstone 1984; Merritt 1983, 1984a,b, 1985, 1988; Miller 1983; Kashlinsky 1987). These studies have typically focused on a subset of the full range of possible galaxy-galaxy and galaxy-background interaction mechanisms. An important goal of much of this work is to develop an understanding of the sequence of morphological types of clusters in terms of dynamical evolution.

A focus of many of these theoretical studies has been investigation of the possible importance of mergers in the production of the cD galaxies found at the centers of many rich, regular clusters. The observation of multiply nucleated first-ranked galaxies was taken as supporting evidence for the galactic cannibalism model by Hausman and Ostriker (1978). The issue of whether the multiple nuclei provide direct evidence of galaxy mergers has been vigorously investigated in recent years, as discussed below.

Cluster Virialization. Peebles (1970) and White (1976b) have performed N-body simulations of evolving galaxy clusters. In this technique, the equations of motion for several hundred to several thousand gravitationally interacting point masses, each representing a galaxy, are directly integrated to follow the evolution of a cluster. This approach is particularly useful for following early evolutionary stages when the cluster has not yet reached virial equilibrium in its inner regions. Peebles (1970) demonstrated the development of the basic core-halo structure seen in rich, regular clusters as a result of gravitational collapse and subsequent dynamical evolution. White (1976b) found that clustering develops in a hierarchical manner with small subclusters forming first and subsequently merging to form rich clusters. On the way to the formation of a single core, the simulated cluster was observed to pass through a binary phase with two central concentrations of galaxies. On the basis of these results, White suggested that the sequence of galaxy cluster morphological types from irregular to regular may represent progressively more advanced stages of dynamical evolution.

Galactic Cannibalism. Hausman and Ostriker (1978) have carried out Monte Carlo simulations of the evolution of galaxies in rich clusters due to the merger process. The galaxies are assumed to populate a dense, virialized core region; the evolution of the cluster structure is not considered. The adopted merger rate is based on the results of Ostriker and Tremaine (1975), who computed the rate of growth of a central galaxy in a cluster due to the infall of less massive galaxies as a result of dynamical friction induced orbit decay. In this computation, the "victim" is assumed to be in circular orbit about the "cannibal." More massive galaxies are affected to a greater degree by dynamical friction and are thus the most likely victims of cannibalism. Hausman and Ostriker (1978) find that the most massive galaxy in a rich cluster tends to grow nearly exponentially in time, due to mergers. Thus the merger process is predicted to run away in dense cluster cores, producing a cD galaxy. This growth must eventually slow as the pool of available "victims" is depleted.

The study of Hausman and Ostriker (1978) is unique in its prediction of observable quantities, making it far more amenable to observational test than other theoretical studies in this area. Their galactic cannibalism model predicts that the brightest galaxies in clusters will systematically differ in their basic properties from other cluster galaxies. Thus, the model predicts bright end modifications of a number of important empirical relations including: (1) the galaxy color-magnitude relation,

(2) the galaxy radius-magnitude relation, and (3) the luminosity function for the cluster.

Merger models provide examples of a "special process" mechanism for the formation of cD galaxies. An alternative view is the statistical picture of Geller and Peebles (1976) in which cD galaxies are just the brightest examples of first-ranked galaxies, and are formed by the same processes that form other elliptical galaxies.

Tidal Stripping and Mergers. Richstone (1976) simulated the evolution of clusters under the action of tidal stripping alone. Richstone and Malumuth (1983) and Malumuth and Richstone (1984) substantially extended this work to also include the effects of dynamical friction in the cluster background and galaxy mergers. Merritt's (1983) simulation of cluster evolution included tidal stripping, two-body relaxation, and dynamical friction in the cluster background. In subsequent work he also included tidal limitation of galaxies by the cluster potential (Merritt 1984a) and accretion of galaxies by the first-ranked galaxy (Merritt 1985). The basic approach used by Richstone and Malumuth is to track the orbits of individual test galaxies in the cluster potential and to use a Monte Carlo treatment of individual galaxy-galaxy interactions. In contrast, Merritt uses a statistical description of the galaxy orbital energy and mass distribution, with a Fokker-Planck treatment of the evolution of orbital energies and masses due to interactions. Thus, Richstone and Malumuth treat the evolution of particular realizations of a statistical ensemble of galaxy clusters, while Merritt treats the mean evolution of the entire ensemble. In principle, these two complementary approaches should produce similar results for a given set of initial conditions.

In the method of Richstone and Malumuth, the full range of orbital eccentricities for galaxies is sampled, unlike the assumption of circular orbits in the work of Ostriker and collaborators. While Merritt's method does not explicitly follow galaxy orbits, his approach gives equal weight to all orbital eccentricities. Thus, the studies of Richstone and Malumuth and of Merritt both provide a test of the circular orbit approximation, which according to Merritt (1985) strongly overestimates the rate of galactic cannibalism.

Richstone and Malumuth (1983) and Malumuth and Richstone (1984) find that tidal stripping significantly reduces most galaxy luminosities in rich, virialized clusters. However, as a large, luminous galaxy at the center of a cluster accretes stars that were tidally removed from other galaxies, its halo can increase in luminosity. In contrast to these results, Merritt (1984a, 1985) finds that no significant amount of tidal stripping occurs due to galaxy-galaxy interactions in rich clusters *following virialization*, i.e. after the cluster formation phase. This is a result of strong truncation of galaxies by the mean tidal field of the cluster in Merritt's simulations, which results in galaxies having much smaller *initial* sizes and thus smaller interaction cross sections, than assumed in other studies. However, Allen and Richstone (1988) argue that mean field tidal limitation of galaxies is less significant than found by Merritt, on the basis of their *N*-body simulations of galaxies orbiting in a cluster potential.

Merritt (1988) has also investigated the truncation of galaxies by the overall cluster potential, by carrying out N-body simulations of galaxies orbiting in a potential dominated by dark matter. He reports similar limits on galaxy sizes ($\lesssim 30\ h^{-1}$ kpc) to the analytic estimates adopted in his earlier cluster evolution simulations. Thus in Merritt's picture, substantial mass loss from galaxies due to tidal effects occurs only during the formation phases of rich clusters rather than continuously throughout their lives. Merritt (1988) notes that the detailed simulations of mass loss in close galaxy-galaxy collisions by Aguilar and White (1985) suggest that collisional mass loss might be an important process in *poor* clusters, which have lower velocity dispersions than rich clusters.

Richstone and Malumuth (1983) and Malumuth and Richstone (1984) find that only a fraction of a set of different Monte Carlo realizations of the same initial conditions produce a cD galaxy, in contrast to the production of a cD in every simulation by Hausman and Ostriker (1978). On the basis of this result, Richstone and Malumuth suggest that statistical fluctuations in the merger process play an important role in determining the present-day morphological type of a cluster. A more extreme result is obtained by Merritt (1984a, 1985), who finds that galaxy mergers do not occur at a significant rate in virialized clusters. As for the case of tidal stripping, Merritt (1985) argues that interactions between galaxies leading to mergers are only important in the original small, low-velocity-dispersion groups of galaxies out of which rich clusters may have formed. Merritt (1984b, 1988) finds, however, that the spatial distribution of galaxies in the central regions of rich clusters will be significantly altered, over the life of the cluster, as a result of dynamical friction due to the intracluster background. This results in the development of a central density cusp, i.e. a density profile that rises continuously with decreasing distance from the cluster center (see also §3.5). There is no significant enhancement of the merger rate in this cusp, since the velocity dispersion remains high. In summarizing his work, Merritt (1988) concludes that mergers probably do play an important role in the formation of cD galaxies, but that these mergers must have occurred before the formation of rich clusters.

3   OBSERVATIONAL STUDIES

A number of observational studies have been carried out in recent years to test the predictions of models for the dynamical evolution of galaxy clusters, particularly those of the galactic cannibalism model of Hausman and Ostriker (1978). A selection of this work is reviewed in the following sections, with an emphasis on my own studies. Other reviews of this subject are given by Dressler (1984) and Tonry (1987).

3.1   Colors of First-Ranked Galaxies

There is a well established correlation between the color of an elliptical galaxy and its total luminosity, in the sense that more luminous galaxies are redder (Visvanathan and Sandage 1977). In the galactic cannibalism model, cD galaxies grow at the expense of their less luminous neighbors; it is therefore predicted that the color of a cD will eventually resemble that of its smaller, bluer victims. (It is assumed that material from the two merging galaxies is well mixed in the merger.) Thus, the galactic cannibalism model predicts that cD galaxies will be bluer than galaxies of similar luminosity that have not grown by accretion (Hausman and Ostriker 1978).

In order to test this prediction of the galactic cannibalism model, I carried out photoelectric photometry of first-ranked galaxies in 26 nearby clusters using the KPNO 1.3 m telescope (Lugger 1984a). The redshift range for this sample is 0.01 – 0.07, corresponding to a distance range of 70 – 400 Mpc. The primary measure of galaxy color adopted in this study is the $U-B$ color index, which is a measurement of the ratio of the luminosity of the galaxy in the near ultraviolet (U) band to that in the blue (B) band. According to the galactic cannibalism model, cD galaxies should have a $U-B$ color index that is about 0.1 magnitude (i.e. about 10%) bluer than nonaccreting galaxies of the same luminosity.

To test for the predicted difference in color, the 26 clusters were divided into two equal sized groups: cD and non-cD. The cD group typically corresponds to BM types I and I-II. A mean color-magnitude [4] relation was fitted to the data, and the mean color residuals from this relation were calculated for the cD and non-cD groups. The mean residuals for the two groups were not found to differ at a statistically significant level. A difference as large as the 0.1 magnitude effect in $U-B$ predicted by the galactic cannibalism model can be ruled out at the $5\,\sigma$ level. This result is illustrated in Figure 1 which shows the color-magnitude diagram for the individual first-ranked galaxies along with the mean color-magnitude relation. The interpretation of this result is discussed in §4.

3.2 *Structure of First-Ranked Galaxies*

Elliptical galaxies follow a well defined radius-magnitude relation over a range of 10^3 in luminosity, in the sense that more luminous galaxies have larger radii (Strom and Strom 1978a,b,c). In the galactic cannibalism model, cD galaxies rapidly swell in radius as they grow in luminosity by accretion. This result is based on the assumptions that escaping stars carry away negligible mass and energy during a merger, and that the merger product is described by the same structural form as the two progenitors. Hausman and Ostriker (1978) predicted that cD galaxies will systematically deviate from the radius-magnitude relation defined by nonaccreting galaxies in the sense of having larger radii at a given luminosity. Observational studies of the structure of first-ranked galaxies, relevant to testing this prediction, have been carried out by Oemler (1976), Hoessel (1980), Schneider, Gunn, and Hoessel (1983), Lugger (1984b), Hoessel and Schneider (1985), and Schombert (1986, 1987).

---

[4] Magnitude is a logarithmic measure of luminosity, with larger magnitudes corresponding to *fainter* luminosities.

Figure 1. Color-magnitude diagram for 25 first-ranked galaxies from Lugger (1984a). cD galaxies are shown as open circles (o) and non-cD galaxies are shown as filled circles (•). Additional surrounding circles indicate those galaxies with the smallest corrections for galactic absorption and the K-effect. The solid line is a linear regression for all of the galaxies.

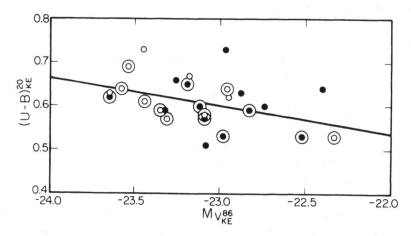

In order to test for the evolution of the radius-magnitude relation predicted by the galactic cannibalism model, I carried out photographic surface photometry for 35 nearby first-ranked galaxies (Lugger 1984b). The distance range is 70 – 600 Mpc. I obtained R-band (central wavelength 6500 Å) plates of these clusters using the Palomar 1.2 m Schmidt telescope and digitized them using the KPNO PDS microdensitometer. Schmidt plates allowed the structure of these galaxies to be studied to large distances from the galaxy centers, ~ 1.5 Mpc for the selected scan size. This is considerably larger than is possible with current generation CCD (charge coupled device) detectors.

The resulting digital pictures were used to determine surface brightness profiles for the galaxies, i.e. brightness as a function of projected distance from the galaxy center. These profiles were fitted with the de Vaucouleurs model,

$$\log S(r) = \log S_e - 3.33 \left[(r/r_e)^{1/4} - 1 \right].$$

The "effective radius" r_e contains half of the total projected light, and thus defines the radial scale of the profile. S_e is the surface brightness at r_e. The de Vaucouleurs model was chosen rather than the Hubble law, which is also commonly used to describe galaxy profiles, since the former is less sensitive to angular resolution limitations. For model fits to galaxy profiles, the de Vaucouleurs effective radius is typically about 11 times larger than the Hubble core radius (Kormendy 1977).

Figure 2 shows de Vaucouleurs model fits to the surface brightness profiles of the

first-ranked galaxies in the clusters A2197 (non-cD) and A1991 (cD). [5] The model
accurately describes the entire profile of the normal galaxy and the inner part of
the profile ($r \lesssim 100$ kpc) of the cD; however, the profile of the cD galaxy lies above
the model at larger radii. This flattening of the outer envelopes of cD galaxies
is a common characteristic and may be taken as a defining property of the class
(Tonry 1987). Schombert (1988) has reviewed possible mechanisms for producing
these diffuse halos, including tidal stripping, cooling flows, mergers, and primordial
origin. It has been suggested that the outer envelope of a cD is a mantle which it
inherits by virtue of its position at the center of the cluster potential well. Fitting
a de Vaucouleurs model provides a means of defining the extent of the first-ranked
galaxy, i.e. determining the location of the galaxy-cluster interface.

Figure 2. R-band surface brightness profiles for the first-ranked
galaxies in the clusters: (a) A2197 and (b) A1991, from Lugger (1982,
1984b). The solid curve is the de Vaucouleurs model that best fits all
of the data. The flat outer part of the profile for A1991 ($r \gtrsim 100$ kpc)
occurs at a level of $\lesssim 1\%$ of the night sky brightness.

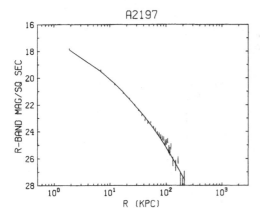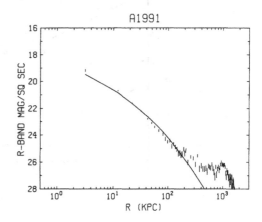

Luminosities were determined for the 35 first-ranked galaxies by integrating the
total light to a range of limiting radii and limiting surface brightnesses. The
dispersion in luminosity, measured to a fixed radius, was found to be only 0.4 mag
for all first-ranked galaxies, consistent with the results of other studies. It is for
this reason that first-ranked galaxies are commonly used as standard candles in
cosmological studies. To test for the predicted deviations from the mean radius-
magnitude relation, clusters were divided into two groups: 11 cD and 24 non-cD.
Within a radius of 50 kpc, the cD first-ranked galaxies were found to be 0.5 mag
($\approx 60\%$) brighter than the non-cD galaxies, at the $5\,\sigma$ level. The mean effective
radius for the cD galaxies (69 kpc) was found to be 80% larger than that for the
non-cD galaxies (38 kpc), at the $3\,\sigma$ level. These results confirm the qualitative
definition that cD galaxies are brighter and larger than non-cD first-ranked

---

[5] The "A" designation indicates a galaxy cluster from the Abell (1958) catalog.

galaxies. A mean radius-magnitude linear regression was fitted to the data for all of the galaxies (see Fig. 3) and the dependence of the radius residuals on cluster type was investigated. No significant dependence was found. First-ranked galaxies of both the cD and non-cD types were found to be consistent with a single radius-magnitude relation. The slope of the relation for first-ranked galaxies is consistent with the range of slopes found by Strom and Strom (1978a,b,c) for fainter elliptical galaxies in clusters. The interpretation of these results is discussed in §4.

Figure 3. Radius-magnitude relation for 35 first-ranked galaxies from Lugger (1984b). cD galaxies are shown as open circles (o) and non-cD galaxies are shown as filled circles (•). The solid line is the linear regression for all of the galaxies. The dashed lines indicate the range of radius-magnitude relations obtained by Strom and Strom (1978a,b,c) extrapolated to the magnitude range of first-ranked galaxies.

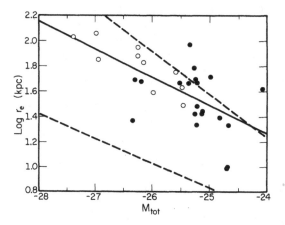

3.3   Luminosity Functions for Galaxy Clusters

Galaxies, both in clusters and in the field, have a characteristic distribution in luminosity ("luminosity function") which is commonly described by the Schechter (1976) parameterization,

$$N(L)\,dL = N^*(L/L^*)^\alpha \exp(-L/L^*)\,d(L/L^*).$$

The parameter N^* is the normalization factor, α is the power-law slope of the faint end of the distribution, and L^* is the "characteristic luminosity," i.e. the luminosity at which the bright end exponential cutoff of the distribution begins to set in. Schechter (1976) determined a standard value of $\alpha = -1.25$ from fits to Oemler's (1974) cluster data. Thus, the number of galaxies per unit luminosity interval *increases* with decreasing luminosity — at least to some faint luminosity

limit. The composite luminosity function for nine Abell clusters is shown in
Figure 4.

Figure 4. Composite cluster luminosity function, excluding first-
ranked galaxies and those second-ranked galaxies that are of the
D (giant) type, from Lugger (1989). Boxes with error bars indicate
the values of the differential luminosity function (number of galaxies
per 0.5 magnitude wide bin). The histogram shows the integrated
luminosity function (number of galaxies brighter than a given
magnitude). Crosses indicate the values of the differential luminosity
function before background correction. The smooth curve through
the values of the differential luminosity function is a Schechter
function fit, with $M^* = -22.70$ and $\alpha = -1.16$.

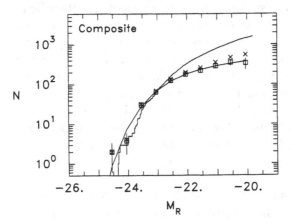

Since both mergers and tidal stripping cause galaxy luminosities to evolve in time,
there will be a corresponding evolution of the luminosity function if these processes
are acting in clusters. Since galactic cannibalism acts most strongly on the most
massive (and thus most luminous) galaxies, it is expected to cause a bright end
depletion of the luminosity function. Hausman and Ostriker (1978) predict a
substantial luminosity function evolution in the sense that the characteristic
luminosity L^* *decreases* in time as a result of accretion of bright galaxies by the
first-ranked galaxy. Tidal stripping is also expected to cause a bright end depletion
of the luminosity function, since larger galaxies have a greater cross section for tidal
stripping.

Observational studies of cluster luminosity functions, relevant to testing these
predictions, have been carried out by a number of investigators including Oemler
(1974), Dressler (1978), Schneider (1982), Godwin, Metcalfe, and Peach (1983,
and references therein), Lugger (1986, 1989), Colless (1989), and Lauer (1989).
Schechter's (1976) analysis of Oemler's (1974) data for 10 clusters indicated a
universal value of L^* with no evidence of significant cluster-to-cluster variation.
Dressler's (1978) analysis of data for 12 clusters, however, found some evidence
for cluster-to-cluster variation of L^* that appeared to be correlated with cluster

morphological type (and thus presumably evolutionary state) in the predicted sense of fainter values of L^* for clusters with more luminous first-ranked galaxies.

In order to investigate evidence for variations of cluster luminosity functions, I carried out photographic photometry for nine Abell clusters using R-band photographic plates that I obtained with the Palomar 1.2 m Schmidt telescope (Lugger 1986). The clusters span a distance range of only a factor of two, 100 – 200 Mpc. Clusters were studied within a uniform size square region 4.6 Mpc on a side, and to a fixed faint end absolute magnitude limit of -19.8 in the R-band. Thus, the sample selection and analysis were carried out in a more uniform manner than in previous studies. Plates were digitized using the KPNO PDS microdensitometer and the digital images were processed with automated object detection and photometry software to determine positions and isophotal magnitudes for cluster galaxies. Star-galaxy discrimination was carried out by plotting isophotal magnitude versus image size; galaxies and stars lie in relatively distinct regions of such a plot. All objects within the galaxy region and near the boundary between the galaxy and star regions were visually inspected under high-power magnification to check the classification. Galaxy magnitudes were used to construct entire cluster luminosity functions. Correction for contamination by foreground galaxies was performed using radial velocity data (when available). Background correction was carried out using the field galaxy counts of Oemler (1974) in Lugger (1986), and using the counts of Butcher and Oemler (1985) in Lugger (1989). Background contamination and thus the background correction increase strongly toward the faint end (see Fig. 4).

Entire cluster luminosity functions were analyzed using both parametric and nonparametric techniques. The Schechter form was fitted to the luminosity function of each cluster and to the composite of all nine clusters. The fits were performed first with both M^* (the absolute magnitude corresponding to the characteristic luminosity L^*) and the faint end slope α allowed to vary, and then with α held fixed at a value of -1.25. The individual values of M^* were then compared with the mean for all clusters. In the nonparametric method, individual luminosity functions were compared to the composite of all nine clusters using the Kolmogorov-Smirnov (K-S) and Wilcoxon rank-sum (W) tests, which are independent of any assumption about the form of the luminosity function. (If the probability that two samples were drawn from the same population was $< 5\%$ according to both tests, the samples were considered to differ significantly.) Due to the rise of the luminosity function toward the faint end, luminosity functions are dominated by galaxies near the faint end limit. Theories of dynamical evolution of clusters, however, predict the greatest evolutionary effects at the bright end. In order to increase the sensitivity of the comparisons to bright end differences, a sliding cutoff magnitude was used to form comparison samples that are increasingly restricted to the bright end of the luminosity function.

The Schechter function parameter values determined from fits to the composite luminosity function with the first-ranked galaxies excluded are $M^* = -22.70 \pm 0.10$ (in the R-band) and $\alpha = -1.16 \pm 0.08$ (Lugger 1989). These are in good agreement with the previous determinations by Schechter (1976) and Dressler (1978). In the comparisons of the individual M^* values (determined with α held fixed) to the mean, a difference as large as $3\,\sigma$ was found only for A569, the cluster with the lowest galactic latitude in the sample and thus with the largest correction for galactic extinction. No significant correlations were found between M^* and any of the following: (1) cluster morphological type, (2) absolute magnitude of the first-ranked galaxy, and (3) cluster central density. The nonparametric tests similarly indicated that the individual luminosity functions are generally consistent with the composite. In a study of the luminosity functions of 14 southern clusters of galaxies, Colless (1989) similarly finds no evidence for significant departures from a universal luminosity function.

3.4 *Intracluster Variation of the Luminosity Function*

In order to more sensitively test for the predicted evolution of the luminosity function as a result of dynamical processes, I carried out an investigation of intracluster variation of the luminosity function (Lugger 1989). The rates for galaxy merger, tidal stripping, and dynamical friction induced orbit decay all increase with the local density of galaxies, since: (1) galaxy-galaxy encounters occur more frequently in denser regions, and (2) the background density increases with galaxy density, increasing the dynamical friction rate. Thus, a variation of the luminosity function between regions of high and low density might be expected if these interaction mechanisms are operating in clusters. In addition, if tidal truncation of galaxies by the mean cluster field is important, as advocated by Merritt (1984a, 1988), then the luminosity function might be expected to depend on the local strength of the tidal field. The tidal field is strongest at about the core radius of a cluster and decreases rapidly with increasing distance outward from the cluster center. Thus, the sizes of galaxies will be most strongly limited in the inner regions. Since all of these interaction mechanisms affect more massive galaxies to a greater degree, the expected sense of the luminosity function variation is a depletion of bright galaxies in high density (inner) regions of clusters relative to low density (outer) regions.

The galactic cannibalism model of Hausman and Ostriker (1978) predicts that the merger process will run away in dense cluster cores, producing a cD galaxy. The subsequent evolution is characterized by growth of the cD by accretion of the surrounding galaxies. Since those galaxies with orbits that pass nearest the cannibal experience the greatest drag, the predicted bright end depletion of the luminosity function would be expected to be seen most strongly within the central region of the cluster. Possible intracluster variation of the luminosity function was not treated by Hausman and Ostriker (1978), who adopted a simplified homogeneous model for clusters in which all of the galaxies are assumed to be "close at hand."

Using the galaxy magnitude and position data base for nine Abell clusters described in §3.3, I separately constructed luminosity functions for inner and outer regions, and high and low density regions (Lugger 1989). These two types of region are the same for clusters with a symmetric decline of galaxy surface density with increasing distance from the center. Six of the nine clusters in the sample have a single, smooth central concentration of galaxies within a radius of ~ 1 Mpc about the center. The other three clusters either have several clumps or an asymmetric central structure. Examples of two clusters with different degrees of clumping are shown in Figure 5.

Figure 5. Galaxy spatial distribution in two clusters, A1656 and A2151 (from Lugger 1982). The contours indicate the local galaxy surface density. The regions shown are 4.32 Mpc on a side. Note the more clumped structure of A2151.

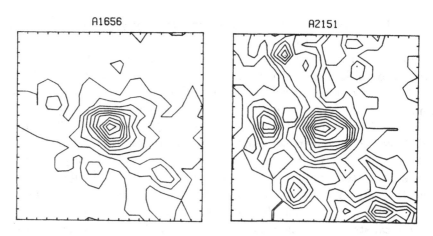

For the radial region comparisons, the standard inner region was a circle of radius 0.5 Mpc about the cluster center; the outer region was a concentric annulus extending from 0.5 to 1.0 Mpc. For the density region comparisons, high and low density samples were constructed by separating galaxies into two groups based on the local galaxy surface density. This was done within square regions of 1.7×1.7 Mpc and 4.6×4.6 Mpc about the cluster center. In addition to constructing high and low density samples for individual clusters, composite high and low density samples were constructed for all nine clusters. Inner/outer and high/low density luminosity function comparisons were carried out using the nonparametric K-S and W tests. A sliding cutoff magnitude was used to separately test for bright- and faint-end differences.

Variations Within Individual Clusters. The radial region comparisons indicate a significant deficit of bright galaxies in the inner region relative to the outer region for three clusters (A1656, A2147, and A2199), according to the nonparametric tests. This effect for A1656 can be seen in Figure 6, which shows the luminosity functions for the inner and outer regions. There is a significant deficit of bright galaxies in the inner region, in this case, even when the two brightest galaxies are included in the inner region sample. The effect becomes stronger when these two galaxies are excluded. The motivation for excluding the brightest galaxies is the prediction by the galactic cannibalism model that cD galaxies will grow at the expense of the other bright galaxies. Quantitative comparison of the inner and outer region luminosity functions (excluding the two brightest galaxies in A1656 and A2147, and the brightest galaxy in A2199) indicates that the observed bright-end deficits correspond to about $(8 \text{ to } 14) \pm 6$ missing galaxies with an associated luminosity of $\sim 14 \pm 7\ L^*$. These luminosity deficits are roughly consistent with the luminosities of the excluded brightest galaxies. Calculation of the average luminosity deficit per galaxy, again excluding the brightest galaxies, indicates that galaxies in the inner region are about 0.3 to 0.5 mag ($\approx$ 30 to 60%) fainter than those in the outer region, with the bright end of the luminosity function primarily responsible for the effect.

Figure 6. Luminosity functions for inner (0.0 – 0.5 Mpc) and outer (0.5 – 1.0 Mpc) regions of A1656. The notation is the same as in Fig. 4 except that no Schechter function fit is shown. The brightest galaxies are included. Note the gap in the inner region luminosity function between the second and third brightest galaxies; this gap is populated in the outer region.

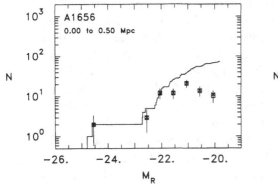 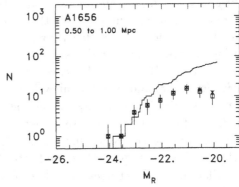

Variations Within the Composite. Nonparametric comparison of the high and low density composite luminosity functions for all nine clusters indicates a significant *excess* of bright galaxies in the high density region when first-ranked galaxies are included. When the first-ranked galaxies are excluded, the luminosity functions are consistent with each other. This indicates that the first-ranked galaxies are not consistent with the luminosity function of the low density region. Schechter function fits to the high and low density samples, including the first-ranked galaxies, give a value of M^* for the high density region that is 0.6 mag ($\approx$ 70%) brighter than that for the low density region, at the $3.3\ \sigma$ level. This effect can be seen in Figure 7.

Figure 7. High and low density composite luminosity functions from Lugger (1989). The notation is the same as in Fig. 4. First-ranked galaxies are included. Note the excess of bright galaxies and the flatter faint end slope in the high density region relative to the low density region.

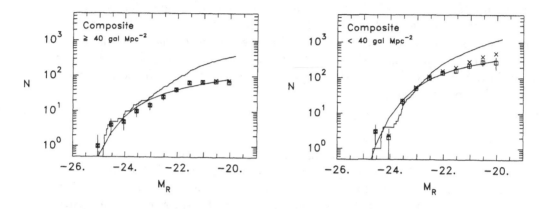

Another significant difference between the high and low density composite luminosity functions, indicated by the nonparametric comparisons, is a flatter faint-end slope for the high density sample. This effect is evident in Figure 7, where the luminosity function for the high density region flattens near the faint end while that for the low density region continues to rise. A flatter faint end in the high density region is also seen for the individual clusters A779, A2197, and A2634 (Lugger 1989). Similar effects were found by Oegerle, Hoessel, and Ernst (1986) for A168, and by Choloniewski and Panek (1987) for E and S0 galaxies from the Center for Astrophysics catalog. Faint-end differences in luminosity functions of cluster galaxies are subject to some uncertainty due to the relatively large background correction for the low density sample. Experiments with different background levels suggest that the flatter faint end in the high density region is not an artifact of incorrect background correction (Lugger 1989). To further investigate this issue, radial velocities are needed for all galaxies in the samples, to enable a direct determination of cluster membership.

3.5 Multiply Nucleated First-Ranked Galaxies

In the galactic cannibalism model, the secondary nuclei observed in about half of all first-ranked galaxies are interpreted as the "undigested remains" of victims that comprise a bound population of satellites that are sinking toward the center of the cannibal on decaying circular orbits. A key prediction of this interpretation is that the velocities of the secondary nuclei with respect to the first-ranked galaxy should be consistent with the inferred circular orbit velocity (~ 300 km s^{-1}) at a distance of $10 - 20$ kpc from the center of the cannibal. Secondary nuclei with substantially larger velocity differences cannot be undergoing significant orbital decay. As discussed below, possible interpretations of these high-speed secondary nuclei are: (1) galaxies that are presently passing close to the first-ranked galaxy on highly elongated orbits, or (2) more distant cluster galaxies that are projected on the first-ranked galaxy.

Tonry (1985a) has investigated the effect of dynamical friction on galaxy orbits in the combined potential of a dominant central galaxy and the cluster in which it resides. His analysis is somewhat similar to that of Merritt (1984b, 1985, 1988), although there are a number of technical differences. Tonry has used his models to interpret velocity data obtained for a sample of multiply nucleated first-ranked galaxies in 14 clusters (Tonry 1985b). His analysis adopts the constraint that the total luminosity accreted by the first-ranked galaxy not exceed 10 L^*, over a 5 Gyr cluster lifetime. Tonry (1985a) finds that only about 30% of the secondary nuclei have velocity differences with respect to the dominant nucleus consistent with bound circular orbits (< 300 km s^{-1}). The velocity distribution extends up to ~ 2000 km s^{-1}. In his model, about 50% of the secondary nuclei have highly elongated orbits and are currently moving at high velocity near pericenter (the point of closest approach to the dominant nucleus). The remaining 20% of the apparent secondary nuclei are interpreted as chance line-of-sight superpositions of more distant cluster galaxies on the central region of the dominant galaxy. Tonry thus concludes that about 70% of the observed multiple nuclei are not in the process of being cannibalized. The preference for elongated orbits with small pericenter distances ($\lesssim 20$ kpc) produces a crowding effect near the cluster center that results in an enhancement in the central density of the cluster. This orbit distribution thus accounts for the large number of secondary nuclei that are observed.

For a sample of 75 "satellites" of first-ranked galaxies in clusters, Cowie and Hu (1986) have analyzed the distribution of velocities relative to the first-ranked galaxies. This sample was constructed by combining new velocity measurements for a set of serendipitously detected galaxies, typically within 100 kpc of the first-ranked galaxies, with previous velocity measurements for secondary nuclei by Tonry (1985b) and by Smith *et al.* (1985). Cowie and Hu find that the distribution of these velocities is best fitted with a two-component Gaussian model in which about 60% of the satellites belong to a population bound to the first-ranked galaxy and the remainder belong to the normal core population of the cluster. The velocity dispersions for these two populations are 250 and 1400 km s^{-1}, respectively. They argue that this two-component model can account for the density enhancements observed around first-ranked galaxies without the predominantly elongated orbits that are required in Tonry's (1984a) model. Cowie and Hu (1986) conclude that the presence of a substantial bound population is qualitatively consistent with the predictions of the galactic cannibalism model.

Bothun and Schombert (1988) have carried out velocity measurements for *complete* samples of galaxies within 400 kpc of the first-ranked galaxy and with $L > 0.5 L^*$, in three cD clusters. The sample sizes range from 9 to 24 galaxies per cluster, including previous velocity measurements. This represents a complementary approach to that of Cowie and Hu (1986), who typically measured velocities for only one or two satellite galaxies per cluster. Bothun and Schombert find that the velocity distributions for A2271 and A2634 are each consistent with a single-component Gaussian, with dispersions of 460 and 840 km s^{-1}, respectively. However, for A2589 there is an excess of low-velocity galaxies ($\lesssim$ 200 km s^{-1}) that may represent a population bound to the first-ranked galaxy. Bothun and Schombert estimate a time scale of $\sim$ 4 Gyr for this bound population to be accreted by the cD as a result of dynamical friction, which would result in an increase of 10 L^* in the luminosity of the cD. They suggest that such accretion occurred at a much earlier time in those cD clusters that do not presently show evidence of a bound population. Bothun and Schombert (1988) also conclude that the orbits of bound galaxies are highly elongated, as in Tonry's (1985a) model. Bothun and Schombert (1989) have extended this analysis to five additional cD clusters, finding clear evidence for a bound population in one of these. When the data for all eight clusters are combined, as in the Cowie and Hu (1986) analysis, Bothun and Schombert also find that the velocity distribution is best fitted with a two-Gaussian model; the dispersions are 120 and 1200 km s^{-1}. However, only 20% of the total galaxy population belongs to the low-velocity-dispersion component, in contrast to the value of 60% found by Cowie and Hu (1986).

Lauer (1988, 1989) has analyzed CCD photometric data for the central regions of 64 clusters of galaxies with multiply nucleated first-ranked galaxies. He assessed evidence for physical interaction between the nuclei and for dynamical evolution of the galaxy distribution in the cluster cores. By attempting to model the first-ranked galaxy as a superposition of noninteracting galaxies, Lauer (1988) was able to detect distortions indicative of tidal interaction. He finds that about half of a sample of 16 multiply nucleated first-ranked galaxies shows evidence of tidal distortion. These distortions are seen both in systems with large velocity differences between the nuclei ($>$ 1000 km s^{-1}) and small velocity differences ($<$ 300 km s^{-1}). Only the low velocity systems are consistent with cannibalism. Lauer (1988) estimates a cannibalism rate of 2 L^* per 5 Gyr, averaged over all Abell clusters, suggesting that accretion cannot build cD galaxies to their present luminosities of $\sim$ 12 L^* during cluster lifetimes.

Using the total luminosity in secondary galaxies within 20 and 100 kpc of the first-ranked galaxy, Lauer (1989) reestimated the amount of accretion that has occurred due to dynamical friction, for different models of the initial spatial distribution of galaxies in the cluster. He finds that the present spatial and luminosity distributions are most consistent with an initial central power-law cusp of slope -1.8, i.e. the initial galaxy density varies with distance r from the cluster center as $r^{-1.8}$. In this case, Lauer finds that the total accreted luminosity is 3 L^* in 10 Gyr. Again, this is considerably smaller than the luminosities of typical cD galaxies. Lauer also finds that the total amount of accreted luminosity depends sensitively on the initial cusp slope. For a slope of -2.0, the total accreted luminosity increases to 7 L^*. However, Lauer finds that the higher cannibalism rate that would be produced by this steeper initial slope would also produce a bright end deficit in the luminosity function for secondary galaxies that he does not observe in his data.

Merrifield and Kent (1989) have investigated the spatial distribution of galaxies within 500 kpc of the central dominant galaxy in 29 clusters. They find an exponential surface density profile, i.e. one that rises toward the center of the cluster, but that does not rise as steeply as a power-law cusp. Beers and Tonry (1986) had earlier found evidence for central power-law cusps in rich clusters, which they took as evidence for Tonry's (1985a) model for the galaxy orbit distribution in the vicinity of a central dominant galaxy. Merrifield and Kent (1989) suggest that the observed exponential central surface density profile is consistent with the action of dynamical friction as calculated by Merritt (1988). No evidence is found for the luminosity segregation that is predicted to result from dynamical friction, but the expected level of segregation is low and therefore may be difficult to detect. Merrifield and Kent find that the velocity distribution observed by Cowie and Hu (1986) may be explained as the result of a background mass distribution that is more extended than the galaxy distribution. This interpretation is consistent with a scenario in which dynamical friction has affected the galaxy distribution. Merrifield and Kent find that projection of galaxies belonging to the exponential cusp, on the central dominant galaxy, can account for at least 50% of the observed secondary nuclei. They further conclude that at least 25% of the observed secondary nuclei may currently be in the process of merger, which is roughly consistent with Lauer's (1988) results for the fraction which both have low velocity and show evidence of physical interaction.

4   DISCUSSION

The observational studies of dynamical evolution of galaxy clusters discussed in §3 have generally been designed to test the galactic cannibalism model of Hausman and Ostriker (1978) which makes the most specific observational predictions. The results are mixed; some of the predictions are generally consistent with the observations, while others are not substantiated.

4.1   Colors of First-Ranked Galaxies

The investigation of the color-magnitude relation discussed in §3.1 indicated that cD galaxies do not significantly deviate from the mean color-magnitude relation defined by all first-ranked galaxies (Lugger 1984a). Thus cD galaxies are not as blue as is predicted by the galactic cannibalism model. However, it may be possible to reconcile this result with a merger picture if the assumption of Hausman and Ostriker (1978) that the color of accreted material remains constant in time is relaxed.

Pickles (1987) reviews theoretical simulations which predict evolution of the metallicities and thus colors of galaxies. [6] In particular, Arimoto and Yoshii (1986) suggest that there may be an *extended* period of star formation in elliptical

---

[6] As the metallicity of a galaxy increases, so do the strengths of the metallic absorption lines that typically lie in the blue region of stellar spectra. This enhanced "line blanketing" depresses the blue emission from the stars in the galaxy, causing the integrated color of the galaxy to become redder.

galaxies, resulting in a steadily increasing metallicity of the intragalactic gas, as the products of nucleosynthesis in successive generations of stars are returned to the interstellar medium. Ultimately, the thermal energy of the gas exceeds its binding energy and the remaining gas escapes from the galaxy, halting further star formation. Since more massive galaxies have deeper gravitational potential wells, the duration of the metallicity enhancement phase should increase with galaxy mass, thus producing a correlation between the luminosity and metallicity of galaxies in the sense observed. If the metallicity enhancement phase extends over a substantial fraction of the life of a cluster, then many — perhaps most — mergers may take place while the metallicities of massive galaxies are still evolving. In this case, the present color of a cannibal would not reflect that of its typical lower mass victim. Instead, the color of a galaxy would be determined by its mass, largely independent of whether or not it is a merger product. In Merritt's (1985) picture, in which mergers can only occur before cluster virialization, enhancement need only last through the cluster collapse phase in order for metallicity evolution to erase the effect of these early mergers on the subsequent color of a galaxy. Thus, the color-magnitude relation may not provide a sensitive test of models in which cD galaxies are built up by successive mergers of less massive galaxies.

4.2 Structure of First-Ranked Galaxies

The investigation of the radius-magnitude relation discussed in §3.2 indicated that cD galaxies do not significantly deviate from the mean relation defined by all first-ranked galaxies. While Hausman and Ostriker (1978) had predicted that cannibal galaxies should have a much steeper radius-magnitude relation than fainter galaxies, it appears that they may have adopted too flat a slope for the unevolved relation, viz. $d\log R/dM = -0.04$. The mean slope of the radius-magnitude relation for elliptical galaxies in clusters found in a number of observational investigations, $d\log R/dM \approx -0.3$ (see Lugger 1984b and references therein), is substantially steeper than this. In fact, this observed slope for elliptical galaxies with a wide range of luminosity is consistent with the value predicted for the *evolved* relation for cannibals by Hausman and Ostriker (1978). Thus, the radius-magnitude relation for first-ranked galaxies may not provide a sensitive test for the action of merger.

4.3 Luminosity Functions

As discussed in §§3.3 and 3.4, analysis of luminosity functions for clusters of galaxies indicated little evidence for significant variations of individual *entire cluster* luminosity functions from a universal form for all clusters (Lugger 1986). However, several forms of *intracluster* luminosity function variation were found (Lugger 1989). Since the effects of dynamical evolution are expected to be most pronounced in the dense central regions of clusters, intracluster comparisons may provide a more sensitive test for these effects than do intercluster comparisons of entire cluster luminosity functions. Effects seen for the central regions of clusters may be washed out when the substantially greater numbers of galaxies in the outer regions are included in the luminosity functions for entire clusters.

The deficits of bright galaxies observed in the central regions of the clusters A1656, A2147, and A2199 (Lugger 1989) may be the result of mergers and/or tidal stripping; both models predict an increasing gap between the luminosity of the first-ranked galaxy and the luminosities of other bright cluster galaxies. The first-ranked galaxy in these three clusters (and the second-ranked galaxies in A1656 and A2147) are of the D (giant) or cD (supergiant) type. A1656, A2147, and A2199 lie in the upper half of the range of central surface density for the nine clusters studied by Lugger (1989), which is consistent with the expectation that merger and tidal stripping rates increase with density. As discussed in §3.4, the luminosity deficits of about $14 \pm 7\ L^*$ in each of A1656, A2147, and A2199 are roughly comparable with the luminosity of the first-ranked galaxy in A2199 and the combined luminosities of the first- and second-ranked galaxies in A1656 and A2147. Thus in terms of luminosity, the D and cD galaxies in these clusters could have been largely produced by accretion of other galaxies. This finding is consistent with both Hausman and Ostriker's prediction that mergers occur throughout the life of a cluster and Merritt's limitation of mergers to the pre-virialization period. The average luminosity deficit of about 0.4 mag per galaxy is also consistent with the luminosity reduction predicted to result from tidal stripping (Malumuth and Richstone 1984). If tidal stripping is primarily responsible for the observed deficit of bright galaxies, then either the stripping rate must increase strongly with luminosity or else the action of tidal stripping must leave the faint end slope of the luminosity function relatively unaffected.

The *excess* of bright galaxies seen in the composite luminosity function for high density regions relative to low density regions, when first-ranked galaxies are included (Lugger 1989), indicates that these first-ranked galaxies are not consistent with the luminosity function for low density regions. Schechter (1976) similarly found that cD galaxies do not fit a composite luminosity function defined by other cluster galaxies. This result is consistent with the "special process" formation mechanism for cD galaxies inherent in the galactic cannibalism and tidal stripping models. Other possible explanations are that: (1) the galaxy formation process may preferentially produce bright galaxies in high density regions, and (2) two-body relaxation and/or dynamical friction have caused massive galaxies to segregate toward the centers of clusters.

The somewhat flatter faint end of the composite luminosity function for high density regions relative to low density regions (Lugger 1989) indicates a greater number of brighter galaxies relative to fainter galaxies in high density regions. Hausman and Ostriker (1978) predict that galactic cannibalism primarily affects the bright end of the luminosity function. Tidal stripping might be expected to have a greater effect than cannibalism on the faint end of the luminosity function, since the stripping rate adopted by Malumuth and Richstone (1984) is appreciable even for faint galaxies. Since tidal stripping can act to either steepen or flatten a power-law luminosity function, more detailed theoretical predictions are necessary before a clear test of tidal stripping as a cause for faint end differences in luminosity functions is possible (Lugger 1989). Another possible cause for a flattening of the faint end of the luminosity function for high density regions is mass segregation, which enhances the relative number of bright galaxies in central regions of clusters. However, Sarazin (1986) notes that observational investigations of the galaxy distribution in clusters provide only weak evidence for mass segregation. This may be due to the tendency of both mergers and tidal stripping to counter mass segregation by reducing the number of bright galaxies in dense, central regions of clusters.

4.4   *Multiply Nucleated First-Ranked Galaxies*

The studies of the structure and/or kinematics of multiply nucleated first-ranked galaxies by Tonry (1985a,b), Cowie and Hu (1986), Bothun and Schombert (1988, 1989), Lauer (1988, 1989), and Merrifield and Kent (1989) generally suggest that galactic cannibalism may be occurring, but at a lower rate than predicted by Hausman and Ostriker (1978). Tonry's (1985a) model is based on the assumption that the total luminosity accreted by the first-ranked galaxy does not exceed 10 L^* over the 5 Gyr life of a cluster. Bothun and Schombert (1988) obtain a similar estimate of the accretion rate, from their data for A2589. Lauer (1989) determines a limit of about 3 L^* for the total accreted luminosity, averaged over all Abell clusters. In comparison, my estimate of the luminosity deficit in each of A1656, A2147, and A2199, which may represent galaxies lost through mergers and/or tidal stripping, is $14 \pm 7\ L^*$. Given that there may be a large range of accretion rates depending on cluster properties, total accreted luminosities $\sim 10\ L^*$ for some clusters may be consistent with Lauer's lower mean value for *all* Abell clusters.

4.5   *Future Work*

Observational investigations of evidence for dynamical evolution of galaxy clusters indicate that the effects of dynamical friction, mergers, and tidal stripping are more subtle than originally predicted. Merritt (1988) has noted that the complex physical nature of a galaxy cluster requires that substantial simplifications be made in computer simulations of its evolution. Different studies have adopted alternative sets of approximations and have, perhaps not surprisingly, arrived at differing results. While a number of investigations suggest that dynamical evolution of clusters has occurred to some degree, the exact amount of evolution is rather uncertain. Thus, more theoretical and observation work remains to be done in order to provide definitive answers.

On the theoretical side, ensembles of simulations with a realistic number of galaxies — of order several thousand — are needed to provide predictions with a greater degree of statistical certainty. A direct assessment of the relative importance of mergers and stripping might be carried out by tracking the evolution of individual bright galaxies. A simulation of the growth of a cD galaxy by accretion, at the center of the cluster potential well, should be performed, taking into account the evolving structure of the cD and the full set of possible orbits in the combined potential of the cD and the cluster.

On the observational side, photometric and spectroscopic data are needed for larger numbers of galaxy clusters in order to increase the statistical certainty of any detected effects. Color-magnitude and radius-magnitude relations should be determined in a uniform way over a wide range of galaxy luminosity to detect possible deviations by first ranked galaxies. Luminosity functions should be determined for more clusters because of the relatively small number of bright galaxies ($L \gtrsim L^*$) in any one cluster. In order to investigate the role of the background correction in the faint-end slope of the luminosity function, radial velocities must be measured for several hundred galaxies per cluster. Such large-scale velocity surveys of cluster galaxies are now practical, due to the advent of high-efficiency, multi-object digital spectroscopy. More photometric and spectroscopic data are particularly needed for the "satellite" galaxy population about first-ranked galaxies, in order to provide firm determinations of the galaxy accretion rate.

With detailed, realistic models for galaxy clusters and an enhanced data base of observations, it should be possible to make much more precise determinations of the degree to which dynamical evolution has shaped the present properties of clusters.

I am indebted to Bill Liller for giving me my start in extragalactic astronomy and for his constant support through the years. I would like to thank Dave Merritt, John Tonry, Steve Kent, and Tod Lauer for informative discussions. I am grateful to Jim Elliot and Dave Merritt for helpful comments on the manuscript.

REFERENCES

Abell, G.O. 1958, *Astrophys. J. Suppl.*, **2**, 311.

Aguilar, L.A. and White, S.D.M. 1985, *Astrophys. J.*, **295**, 374.

Allen, A.J. and Richstone, D.O. 1988, *Astrophys. J.*, **325**, 583.

Arimoto, N. and Yoshii, Y. 1986, *Astron. Astrophys.*, **164**, 260.

Bautz, L.P. and Morgan, W.W. 1970, *Astrophys. J. Lett.*, **162**, L149.

Beers, T.C. and Tonry, J.L. 1986, *Astrophys. J.*, **300**, 557.

Binney, J. and Tremaine, S. 1988, *Galactic Dynamics*, (Princeton: Princeton University Press), §7.1.

Bothun, G.D. and Schombert, J.M. 1988, *Astrophys. J.*, **335**, 617.

————. 1989, preprint.

Butcher, H.R. and Oemler, A., Jr. 1985, *Astrophys. J. Suppl.*, **57**, 665.

Choloniewski, J., and Panek, M. 1987, in *IAU Symposium No. 127, Structure and Dynamics of Elliptical Galaxies*, ed. T. de Zeeuw, (Dordrecht: Reidel), p. 457.

Colless, M. 1989, *Mon. Not. Roy. Astron. Soc.*, **237**, 799.

Cowie, L.L. and Hu, E.M. 1986, *Astrophys. J. Lett.*, **305**, L39.

Dekel, A., Lecar, M., and Shaham, J. 1980, *Astrophys. J.*, **241**, 946.

Dressler, A. *Astrophys. J.*, **223**, 765.

————. 1984, *Ann. Rev. Astron. Astrophys.*, **22**, 185.

Gallagher, J.S. and Ostriker, J.P. 1972, *Astron. J.*, **77**, 288.

Geller, M.J. and Peebles, P.J.E. 1976, *Astrophys. J.*, **206**, 939.

Godwin, J.G., Metcalfe, N., and Peach, J.V. 1983, *Mon. Not. Roy. Astron. Soc.*, **202**, 113.

Hausman, M.A. and Ostriker, J.P. 1978, *Astrophys. J.*, **224**, 320.

Hoessel, J.G. 1980, *Astrophys. J.*, **241**, 493.

Hoessel, J.G. and Schneider, D.P. 1985, *Astron. J.*, **90**, 1648.

Kashlinsky, A. 1987, *Astrophys. J.*, **312**, 497.

Kormendy, J. 1977, *Astron. J.*, **218**, 333.

Lauer, T.R. 1988, *Astrophys. J.*, **325**, 49.

————. 1989, *Astrophys. J.*, submitted.

Lecar, M. 1975, in IAU Symposium No. 69: Dynamics of Stellar Systems, ed. A. Hayli, (Dordrecht: Reidel), p. 161.

Lugger, P.M. 1978, *Astrophys. J.*, **221**, 745.

————. 1982, Ph.D. Thesis, Harvard University.

————. 1984a, *Astrophys. J.*, **278**, 51.

————. 1984b, *Astrophys. J.*, **286**, 106.

————. 1986, *Astrophys. J.*, **303**, 535.

————. 1989, *Astrophys. J.*, **343**, 572.

Malumuth, E.M. and Richstone, D.O. 1984, *Astrophys. J.*, **276**, 413.

Mathews, T.A., Morgan, W.W., and Schmidt, M. 1964, *Astrophys. J.*, **140**, 35.

Merrifield, M.R. and Kent, S.M. 1989, *Astron. J.*, **98**, 351.

Merritt, D. 1983, *Astrophys. J.*, **264**, 24.

————. 1984a, *Astrophys. J.*, **276**, 26.

————. 1984b, *Astrophys. J. Lett.*, **280**, L5.

————. 1985, *Astrophys. J.*, **289**, 18.

————. 1988, in *The Minnesota Lectures on Clusters of Galaxies and Large-Scale Structure*, ed. J.M. Dickey, (Provo: Brigham University Press), p. 175.

Miller, G.E. 1983, *Astrophys. J.*, **268**, 495.

Oegerle, W.R., Hoessel, J.G., and Ernst, R.M. 1986, *Astron. J.*, **91**, 697.

Oemler, A., Jr. 1974, *Astrophys. J.*, **194**, 1.

————. 1976, *Astrophys. J.*, **209**, 693.

Ostriker, J.P. and Tremaine, S.D. 1975, *Astrophys. J. Lett.*, **202**, L113.

Peebles, P.J.E. 1970, *Astron. J.*, **75**, 13.

Pickles, A. 1987, in *IAU Symposium No. 127, Structure and Dynamics of Elliptical Galaxies,* ed. T. de Zeeuw, (Dordrecht: Reidel), p. 203.

Richstone, D.O. 1975, *Astrophys. J.*, **200**, 535.

————. 1976, *Astrophys. J.*, **204**, 642.

Richstone, D.O. and Malumuth, E.M. 1983, *Astrophys. J.*, **268**, 30.

Sarazin, C.L. 1986, *Rev. Mod. Phys.*, **58**, 1.

————. 1988, *X-ray Emission from Clusters of Galaxies*, (Cambridge: Cambridge University Press).

Schechter, P. 1976, *Astrophys. J.*, **203**, 297.

Schneider, D.P. 1982, Ph.D. Thesis, California Institute of Technology.

Schneider, D.P., Gunn, J.E., and Hoessel, J.G. 1983, *Astrophys. J.*, **268**, 476.

Schombert, J.M. 1986, *Astrophys. J. Suppl.*, **60**, 603.

————. 1987, *Astrophys. J. Suppl.*, **64**, 643.

_____. 1988, *Astrophys. J.*, **328**, 475.

Smith, R.M., Efstathiou, G., Ellis, R.S., Frenk, C.S., and Valentijn, E.A. 1985, *Mon. Not. Roy. Astron. Soc.*, **216**, 71P.

Spitzer, L. 1958, *Astrophys. J.*, **127**, 17.

Strom, K.M. and Strom, S.E. 1978a, *Astron. J.*, **83**, 73.

_____. 1978b, *Astron. J.*, **83**, 1293.

Strom, S.E. and Strom, K.E. 1978c, *Astron. J.*, **83**, 732.

Tonry, J.L. 1985a, *Astrophys. J.*, **291**, 45.

_____. 1985b, *Astron. J.*, **90**, 2431.

_____. 1987, in *IAU Symposium No. 127, Structure and Dynamics of Elliptical Galaxies,* ed. T. de Zeeuw, (Dordrecht: Reidel), p. 89.

Tremaine, S. 1981, in *The Structure and Evolution of Normal Galaxies*, eds. S.M. Fall and D. Lynden-Bell, (Cambridge: Cambridge University Press), p. 67.

Visvanathan, N. and Sandage, A. 1977, *Astrophys. J.*, **216**, 214.

White, S.D.M. 1976a, *Mon. Not. Roy. Astron. Soc.*, **174**, 19.

_____. 1976b, *Mon. Not. Roy. Astron. Soc.*, **177**, 717.

QUESTION AND ANSWER

S. Strom: Can the bright end deficits of the luminosity function that you find for the inner regions of A1656, A2147, and A2199 be accounted for by the amount of tidal stripping that we find in central regions of clusters from our studies of the radius-magnitude relation?

P. Lugger: Yes. The bright end deficit of the luminosity function corresponds to an average luminosity deficit of about 0.4 mag per galaxy within the inner regions of A1656, A2147, and A2199. This deficit is consistent with the prediction of the tidal stripping model of Malumuth and Richstone (1984), and with your observational finding that elliptical galaxies in high density regions of clusters have decreased by 0.5 mag in luminosity relative to those in low density regions. This issue is discussed in more detail in §§3.4 and 4.3 of the present article and in Lugger (1989).

WHAT IS IN THE X-RAY SKY?

R.E. Schild
Harvard-Smithsonian Center for Astrophysics
Cambridge, MA 02138 USA

Abstract. Optical identifications of hundreds of sources from the Einstein Observatory Medium Survey are reviewed. Identifications include stars, active galactic nuclei (AGN's)[1], and galaxy clusters. Many of the images of AGN's show "fuzz" which in many cases is not symmetrical around the stellar object, indicating that galaxy interactions may be responsible for the activity in many AGN's.

1   INTRODUCTION

Bill, the identification of X-ray sources held your interest throughout the 1970's. Building on this work, a really new understanding of QSO's seems to be emerging in the 1980's. Probably most of us remember the days when X-ray source positions were coming out of the American Science and Engineering group in the early 1970's and your knowledge of catalogues, plate stacks, and the entire fabric of optical astronomy allowed you to become the guru of optical identifications of X-ray sources. I am personally reminded of your contributions whenever we work on the Einstein Observatory Deep Survey X-ray sources; the first thing we always do is look at the beautiful plates of the Ursa Minor and Draco fields that you took at the KPNO 4 meter telescope.

I'm sure that you can remember the days when the Einstein Observatory was being prepared for launch. The *Uhuru* satellite had given us a bewildering collection of sources such as Sco X-1 and Her X-1, but it was known that the X-ray background seen at the galactic poles could not come from such classes of objects confined to the galactic plane. It has turned out that the Einstein Observatory Deep Survey has not been so helpful in identifying the origins of the background seen at high galactic latitude because the areal coverage is too small and the objects discovered are too faint for ground-based observation and study. The Einstein Medium Survey of serendipitously discovered X-ray sources has been more useful because of its larger sky coverage, in spite of its lower sensitivity.

---

[1] In this article, the term "active galactic nuclei" refers to both the nuclei of Seyfert galaxies and to quasi-stellar objects (QSO's = quasars).

2 NEW RESULTS

So what has the Einstein Observatory Medium Survey found at the galactic poles? A program of complete identifications for 112 X-ray sources has shown that: 25% are galactic stars, 16% are clusters or groups of galaxies, 8% are BL Lac's or normal galaxies, and 51% are AGN's. Thus, it now appears that the program to identify the source of the X-ray background will require an understanding of the evolution of X-ray emission in clusters of galaxies and AGN's.

Right now the most interesting progress seems to be coming from our studies of the AGN's. We are finding many AGN's which are not currently in catalogues of QSO's and this is at least in part because they have been discriminated against in surveys to date that search for objects having blue color. Whereas most QSO searches are for objects having B−V < 0.6, about a third of our X-ray selected AGN's have colors redder than this limit. We presume that this means that X-ray selected AGN's frequently lack the strong non-stellar blue continuum characteristic of "real" QSO's and they may be more closely related to Seyfert galaxies. We have found that most of the red X-ray identifications have luminosities fainter than a typical brightest cluster elliptical galaxy and most have a resolved, fuzzy appearance on optical images taken in good seeing.

Something quite unexpected that we are finding is that many of the AGN images are not symmetrical. We have found this to be the case both for the BL Lac's and for the QSO/Seyferts. The image structures seen are suggestive of galaxy interaction. As the group studying Medium Survey identifications compares notes, it has found it useful to describe the morphology of galaxies observed in $1''$ seeing on the following classification scheme:

Type Q1 is stellar and type Q2 is a stellar object with fuzz. If the fuzz is not symmetrical around the stellar core, the object is Q3 and is presumed to be an interacting system. Type Q4 is distinguished by the presence of a secondary nucleus in a common envelope, and Q5 is assigned if the $V = 25th$ magnitude per square arcsec isophote is pinched ("peanut shaped"). We frequently use Q3+ to denote all of the form types which appear to indicate a galaxy interaction.

Figures 1–3 show some examples of AGN morphology found in the Medium Survey; the asymmetries seen in Figures 2 and 3 appear to indicate interaction. All are from 10 or 20 minute CCD (charge coupled device) exposures taken with a red filter at the f/13.5 focus of the Mt. Hopkins 61 cm telescope. The scale on our RCA CCD is 0.73″/pixel. North is up and east is to the left in all illustrations.

Thus far, 21 BL Lac sources in the Expanded Medium Survey have been identified from the strength of their radio emission and featureless spectra. Many of them have been resolved and are classified with the following statistics:

Class:	Q1	Q2	Q3+	Unclassified	Total
# objects	7	4	6	4	21
% of classified	41%	24%	35%		

We are currently monitoring the brightnesses of the 21 new BL Lac objects. Most show only minor brightness fluctuations of a few tenths of a magnitude over the typically two years of brightness monitoring.

Figure 1. The AGN 1306.7-0121 has a redshift of $z = 0.08$ and is left of center in this 10 minute exposure. The image is clearly extended, but the isophotes traced by computer do not show evidence of asymmetry indicative of interaction; the galaxy is classified Q2.

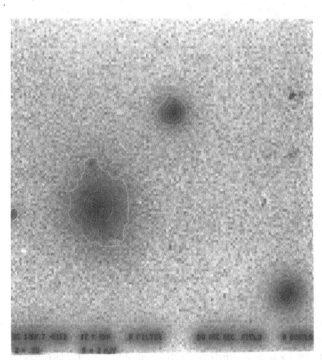

The expanded Medium Survey is expected to yield 400-500 AGN's but so far only 130 have been classified from their images taken in 1–2″ seeing. The statistics of the classifications made so far are:

Class:	Q1	Q2	Q3+	Total
# objects	6	31	28	65
% of total	9%	48%	43%	100%

We see from the statistics above and from statistics of the BL Lac's a common result: the number of objects classified Q3 and higher, the ones which look like interacting galaxies, number about 40% of the class as a whole. This result is in striking agreement with Hutchings and Campbell (1983), who found that at least 30% of QSO's show evidence of galaxy-galaxy interaction.

Figure 2. The AGN 1420.1+2556 is shown here with contours from a 20 minute exposure. The AGN is just below center and clearly has asymmetrical contours.

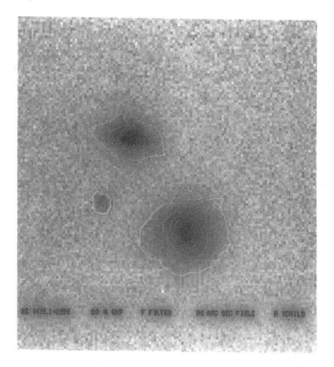

3 DISCUSSION

So what new understanding has emerged from our studies of X-ray sources? In addition to the statistics of what fraction are stars, AGN's, and galaxy clusters, we are coming to a new understanding of AGN's. In the first place, we seem to find that there is a continuous range in activity from Seyfert to QSO, and this is probably related to the rate of fueling of the nuclear condensed object. The results of optical identifications of IRAS (Infrared Astronomical Satellite) sources suggest that most extragalactic IR sources are interacting galaxies, but only a few are AGN's. At the same time, about a third of the most luminous AGN's are seen to be interacting galaxies, and presumably many more are interactions where we do not see evidence of the disturbance because the intruding galaxy is swallowed up or is too far or too faint or something. My conjecture is that you get a QSO only when a massive black hole at the center of a galaxy is fueled by a galaxy encounter. A black hole in a galaxy not undergoing encounter probably is fueled by the galaxy's interstellar medium and a Seyfert galaxy results. You would then expect the rate of fuel deposition to vary widely from object to object and also to vary widely in time to produce a broad range of nuclear luminosities in AGN's, as observed.

Figure 3. The X-ray source 0735.6+7421 lies at the center of a cluster at a redshift of $z = 0.216$. The dominant image contains three nuclear condensations. In the image with reduced grey scale levels shown here, a luminous bridge connecting with two additional bright nuclear condensations is evident. The source is classified Q5. This image covers a field of 90''.

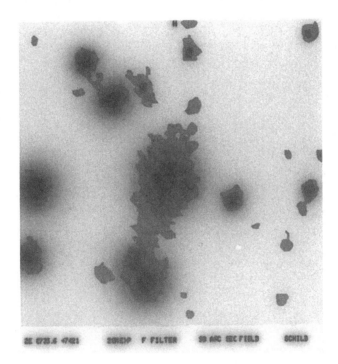

REFERENCES

Hutchings, J.B. and Campbell, B. 1983, *Nature*, **303**, 584.

QUESTION AND ANSWER

M. Mattei: How do you compensate for image smear on an altitude-azimuth mount telescope such as the Multiple Mirror Telescope (MMT)?

R. Schild: All MMT instruments are mounted to a mechanical image rotator at the focal plane of the telescope. This device rotates at a rate computed by the telescope control computer appropriate to the altitude and azimuth of telescope pointing. An elaborate system of cable wraps is needed to get signals to and from the focal plane instruments.

EINSTEIN DEEP SURVEYS

S. Murray, C. Jones, and W. Forman
Harvard/Smithsonian Center for Astrophysics, 60 Garden St.,
Cambridge, MA

ABSTRACT. In this contribution we summarize the present state of our knowledge of the faintest extragalactic sources and their contribution to the diffuse extragalactic x-ray background (XRB) derived from the Einstein Deep and Medium Surveys. Unlike the earlier OSO, UHURU, and Ariel surveys, the extragalactic sources detected in the Einstein surveys are predominantly active galactic nuclei (AGN: quasars and Seyfert galaxies). The surveys show that the AGN population undergoes considerable evolution and that at early epochs AGN produce more emission per unit comoving volume than at the present. Clusters of galaxies show no significant evolution based on their number-flux relation derived from the surveys. Integrating the contribution of all the discrete extragalactic sources, the Einstein surveys detect discrete sources faint enough to comprise $\approx$20-30% of the XRB. We describe Bill Liller's earliest contributions to selecting candidate optical counterparts of these faintest of known x-ray sources.

INTRODUCTION

The Einstein Observatory was launched on 13 November 1978 with the High Resolution Imager (HRI) at the focus of the telescope (see Giacconi et al. 1979a for details of the instrument and telescope). The first phase of operation of the observatory consisted of two weeks of "activation" for each of the four focal plane instruments during which time the actual operating characteristics of the instruments and the telescope were determined. During this activation phase scientific observations were planned and carried out that had great promise of new discoveries. One of these observations was a deep survey the goal of which was to probe the x-ray sky with the greatest possible sensitivity. A week after launch on 20 November 1978, a deep survey was begun in the constellation Ursa Minor at $14^h14^m23^s$ and $73°10'$. This initial survey lasted 29,208 seconds and achieved a sensitivity of 2×10^{-13} ergs cm^{-2}s^{-1} (5σ limit on detection). This first survey undertaken with the Einstein Observatory was capable of detecting sources approximately 100 times fainter than was any previous x-ray survey and demonstrated the full power of imaging observations for x-ray astronomy.

We begin this contribution with a discussion of the number-flux relation and the discrete source contribution to the XRB. Next, we describe Bill Liller's early successes at identifying faint Einstein sources and we conclude with a brief description of the capabilities of the High Resolution Camera on the Advanced X-ray Astrophysics Facility (AXAF) for resolving the x-ray background.

The primary goal of the deep surveys was to investigate the nature of the XRB by directly imaging, counting and identifying faint sources. This is the most direct procedure for determining the contribution of discrete sources to the XRB.

The presence of an x-ray background was discovered by the 1962 rocket experiment that found Sco X-1 (Giacconi et al. 1962). The all-sky observations carried out with the UHURU satellite were used to establish that the XRB was isotropic to within a few percent. To the level of UHURU's and Ariel V's limiting sensitivity ($\sim 10^{-11}$ ergs cm^{-2}s^{-1}), the number-flux relation (log N-log S) was found to be consistent with that expected from a uniform distribution of sources in a non-evolving Euclidean Universe (Matilsky et al. 1973; Warwick and Pye 1978). The extragalactic x-ray sources found by these early x-ray satellites were evenly divided between rich clusters of galaxies and AGN, particularly, Seyfert galaxies.

In addition to the survey in Ursa Minor, surveys were carried out in Cetus, Draco, Eridanus, Canes Venatici, and Pavo. A total of eight deep survey fields were observed with the Einstein Imaging Proportional Counter (IPC) with exposure times ranging from 28,000 to 74,000 seconds. Limiting sensitivities reached $\approx 5 \times 10^{-14}$ ergs cm^{-2}s^{-1}, nearly 1000 times fainter than pre-Einstein surveys. High Resolution Imager (HRI) observations generally were used to map the inner parts of the IPC fields.

LOG N-LOG S AND THE DISCRETE SOURCE CONTRIBUTIONS TO THE X-RAY BACKGROUND

The extension of the number-flux distribution (log N-log S) of extragalactic sources to fluxes nearly 1000 times smaller than those available from the first x-ray surveys of OSO-8, UHURU, and Ariel 5 have allowed investigators to study the cosmological evolution of x-ray sources and to estimate their contributions to the x-ray background.

Integrating the number-flux relation to the faintest flux limits reached in the Einstein observations yields 7.0×10^{-9} erg cm^{-2}s^{-1}Sr^{-1} in the 1-3 keV band. This compares to the extragalactic x-ray background (1-3 keV) of 1.9×10^{-8} erg cm^{-2}s^{-1}Sr^{-1}. Thus, about 35% of the extragalactic x-ray background in the 1-3 keV band has been resolved into discrete sources (Giacconi et al. 1979b, Murray 1981, Griffiths et al. 1983). The most recent analysis of the Einstein deep survey (Primini et al. 1988) used only those sources detected at $\geq 4.5\sigma$ and placed a lower limit on the discrete source contribution to the x-ray background of $\sim$ 20%. Unlike the less sensitive UHURU and Ariel V surveys which found primarily clusters of galaxies and Seyfert galaxies, the Einstein survey found that at low flux levels the predominant contributors to the XRB are quasars and Seyfert galaxies. Studies of complete samples of AGN show that, with the allowed luminosity and evolution function, AGN account for at least 50% of the x-ray background measured at 2 keV (Maccacaro, Gioia, and Stocke 1984) and may produce 100% of the XRB (Marshall et al. 1984). These large inferred contributions are based on the extrapolation of the x-ray properties of AGN to fluxes fainter than the current Einstein limits.

Although the analysis of the early UHURU, Ariel 5, and HEAO-1 surveys suggested that the number-flux relation could be described as arising from a uniform source distribution in a non-evolving Euclidean Universe, the more sensitive Einstein surveys have shown that this result may be due to offsetting effects in the different source populations. In particular, from the Einstein Medium Sensitivity Survey, Gioia et al. (1984) have found that AGN have a steeper number-flux distribution than that expected for a non-evolving, uniformly distributed population in a Euclidean universe. This result is consistent with the suggestion that there were more, or brighter, AGN at earlier cosmological epochs. Also, the number-flux distribution of clusters of galaxies is flatter than Euclidean, which is at least partially attributable to cosmological expansion.

EARLY WORK ON OPTICAL IDENTIFICATIONS

Following the analysis of the x-ray data — generating images, detecting and locating sources, and determining their fluxes — the major task was the optical identification of the sources to learn the nature of these extremely faint objects. It was in this identification process that Bill Liller applied his talents as an optical astronomer.

Following his interests of identifying x-ray sources, begun using positions from the UHURU Catalogs (e.g., Giacconi et al. 1972, Forman et al. 1978), Bill Liller actively began to identify the new faint Einstein sources. Figure 1 shows an example of Bill's work (and notes) on a source at $\alpha = 14^h12^m31^s.6$, $\delta = 73°03'45''$ dated 5 January 1979, only six weeks after the Einstein observations were taken (and before the full production data had been analyzed). The final production processing for this field in Ursa Minor had an updated position for this source of $\alpha = 14^h12^m33^s.3$, $\delta = 73°03'48''$ and a source intensity of $3.6 \pm 0.6 \times 10^{-3}$ HRI counts s^{-1}. The source flux was found to be 2.7×10^{-13} ergs s^{-1}.

For the selection of candidates for this source identification Bill Liller used the Palomar Sky Survey and two astrometric plates from the Harvard collection (taken in 1934 and 1936) to establish the UV excess of a faint candidate, $m_B \approx 18.8$, near the center of the HRI error circle. As Figure 1 shows, Bill obtained two additional plates of the field, an unfiltered image (21 minute exposure by R. McCrosky) and a U plate (by J. Bulger and G. Schwarz) with an exposure time of almost 4 hours from Agassiz! No sources were too faint for him to go after. Subsequent spectroscopic observations of this UV bright candidate showed it to be a quasar with z=0.43 (Primini et al. 1988). Although extensive optical and radio observations had been made in preparation for identifying deep survey sources including Westerbork radio maps, and Kitt Peak 4-meter multicolor plates by others, Bill successfully forged ahead using more modest facilities.

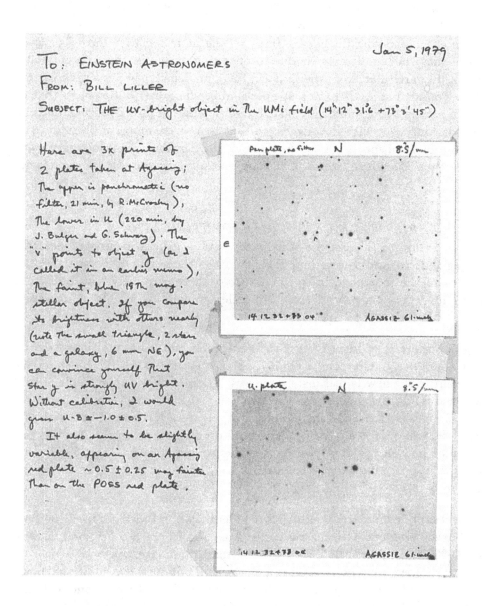

Figure 1: This copy of the finding charts and notes from Bill Liller shows the selection of a candidate for a Deep Survey source identification.

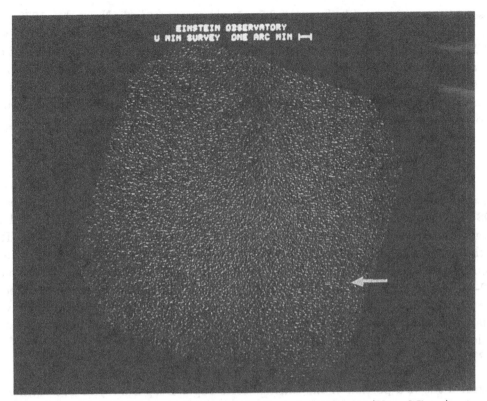

Figure 2: The Einstein image of the first deep survey (Ursa Minor).

The x-ray image of this first survey is shown in Figure 2 which contains the source (see arrows) whose optical ID is discussed in Figure 1. The Ursa Minor region of the sky containing this quasar was re-observed with the HRI on 7-12 June 1979 (six months after the first observations). For six days the HRI was used to obtain the most sensitive x-ray survey carried out by the Einstein Observatory having a total exposure time of 331,123 seconds and a limiting flux of 1.3×10^{-14} ergs cm^{-2}s^{-1} (5σ limit). We refer to this observation as the Super Deep Survey (SDS — Einstein sequence 4279). Figure 3a shows a portion of the image resulting from this survey.

Four faint sources were detected with count rates ranging from 0.50 - 1.23 $\times 10^{-3}$ counts s^{-1} and fluxes of 3.8 - 9.2 $\times 10^{-14}$ ergs cm^{-2}s^{-1}. The strongest source in this field is located at the same coordinates (within the errors) as the source identified by Bill Liller. However, the source count rates from the two observations differ. The first survey gave a count rate of $3.6 \pm 0.6 \times 10^{-3}$ counts s^{-1} while the second gave $1.23 \pm 0.12 \times 10^{-3}$ counts s^{-1} which formally differ at the 3.8σ level. This would suggest that the source varies by factors of a few on a timescale of ≈ 6 months. The two intensity measurements are subject to systematic uncertainties which need to be fully investigated before source variability can be confirmed. However, the change in intensity seems larger than can be accounted for by those effects with which we are familiar. Furthermore, the IPC was used to observe this same region on 24-26 September 1979 (3 months after the second HRI observation). The flux at that time was $5.4 \pm 1.0 \times 10^{-14}$ ergs cm^{-2}s^{-1} (Primini et al. 1988). The HRI observations correspond to fluxes of 2.7×10^{-13} ergs cm^{-2}s^{-1} (November 1978) and 9.2×10^{-14} ergs cm^{-2}s^{-1} (June 1979). Although direct comparisons of HRI and IPC fluxes are subject to large systematic errors due to our lack of knowledge of the source spectrum, these observations suggest a decline by a factor of five in intensity for this source over a timescale of about one year. Although variability of quasars is a common phenomenon at x-ray wavelengths, the magnitude of the variation described above is larger than commonly observed (see Zamorani et al. 1984).

FUTURE OBSERVATIONS

The Einstein Deep Surveys have imaged about 30% of the XRB. Future missions such as AXAF with the High Resolution Camera (HRC; see Murray et al. 1987 for detailed description), an improved version of the Einstein HRI, will have larger effective area and improved quantum efficiency at higher energies. The HRC with 0.5 arc second resolution will be able to detect sources nearly 50 times fainter (limiting 5σ sensitivity of 2×10^{-16} ergs cm^{-2}s^{-1} in 3×10^5 second exposure in the band 0.2-4.0 keV or 10^{-15} ergs cm^{-2}s^{-1} in the band 0.2-8.0 keV) than could be detected in the Einstein Super Deep Survey and should make a major contribution to resolving a large fraction of the XRB. An extrapolation of the observed number-flux relation to lower fluxes predicts that, at a flux of 1.7×10^{-15} ergs cm^{-2}s^{-1}, 100 percent of the XRB would be resolved into discrete sources.

Figure 3 shows a portion of the Einstein HRI Super Deep Survey in Ursa Minor. A simulation of the AXAF HRC image of this field is shown on the right.

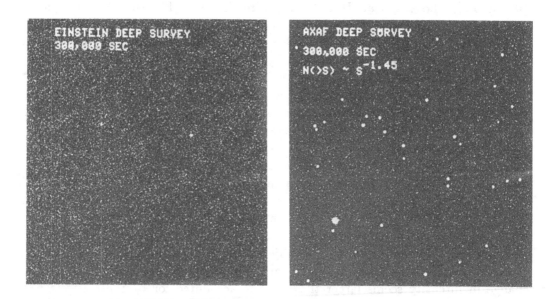

Figure 3: The Einstein image of the super deep survey (left) and a simulation of the same field observed by AXAF (right).

In producing this simulation, we have assumed that the number-source flux relation can be extrapolated to faint limits such that all of the XRB is resolved into discrete sources. While AGN are the predominant sources of x-ray emission, approximately 25% of the sources are assumed to be stars in our Galaxy, consistent with the earlier and less sensitive Einstein deep surveys and medium sensitivity surveys. Note that in computing the contribution to the XRB (defined here as the extragalactic x-ray background), the stellar sources are ignored, since they do not contribute significantly at energies above a few keV.

The most often discussed alternative to discrete sources as the cause of the XRB is a hot, low density gas filling intergalactic space and producing a truly diffuse x-ray background (e.g., Marshall et al. 1980). Due to the cosmological importance of such a hot component of the Universe, the AXAF observations will be particularly important in limiting or finding this diffuse gas.

Acknowledgements

We acknowledge the care and expertise of K. Modestino in preparing this manuscript and of J. Lupo in preparing figures 2 and 3. This work was supported through NASA Contract NAS8-30751.

REFERENCES

Forman, W., Jones, C., Cominsky, L., Julien, P., Murray, S.,
        Peters, G., Tananbaum, H., Giacconi, R. 1978, Ap J. Supp. 38. 357.

Giacconi, R., Gursky, H., Paolini, F.R., and Rossi, B.B. 1962
        Phys. Rev. Letters, 9, 439.

Giacconi, R., Murray, S., Gursky, H., Kellogg, E., Schreier, E.,
        and Tananbaum, H. 1972, Ap.J., 179, L81.

Giacconi, R., et al. 1979a, Ap.J., 230, 540.

Giacconi, R., et al. 1979b, Ap.J., 234, L1.

Gioia, I., Maccacaro, T., Schild, R., Stocke, J., Liebert, J.,
        Danziger, I., Kunth, D., and Lub, J. 1984, Ap.J., 283, 495.

Griffiths, R. et al. 1983, Ap. J. 269, 375.

Maccacaro, T., Gioia, I.M., and Stocke, J.T. 1984, Ap. J., 283, 486.

Marshall, F.E., et al. 1980, Ap.J., 235, 4.

Marshall, H., Avni, Y., Braccesi, A., Huchra, J.P., Tananbaum, H.,
        Zamorani, G., and Zitelli, V. 1984, Ap.J., 283, 50.

Matilsky, T., Gursky, H., Kellogg, E., Tananbaum, H., Murray, S.,
        and Giacconi, R. 1973, Ap.J., 181, 753.

Murray, S.S. 1981, in X-ray Astronomy with the Einstein Satellite
        (ed. R. Giacconi: D. Reidel) p.281.

Murray, S.S., et al. 1987, Astro. Lett. and Communications, 26, 113.

Primini, F., Murray, S.S., Burg, R., Huchra, J., and Schild, R.
        1988, preprint.

Warwick, R.S. and Pye, J.P. 1978, MNRAS, 183, 169.

Zamorani, G., Giommi, P., Maccacaro, T., and Tananbaum, H. 1984,
        Ap. J., 278, 28.

History, Lore, and Archaeoastronomy

ROBERT WHEELER WILLSON: HIS LIFE AND LEGACY

B. L. Welther
Harvard-Smithsonian Center for Astrophysics,Cambridge, MA 02138

Abstract. For the 20 years between 1962 and 1982, William Liller held the Robert Wheeler Willson Professorship of Applied Astronomy at Harvard University. Therefore, the Liller Symposium presents an appropriate occasion to remember Willson and his contribution to astronomy at Harvard.

From the seventeenth through the nineteenth centuries Harvard usually offered a course in mathematics or engineering which included some practical aspects of astronomy applicable to navigation and surveying. However, it was not until the twentieth century that courses and teaching in astronomy *per se* blossomed and flourished at Harvard. Because Harlow Shapley developed a Ph.D. program at the Observatory in the 1920s, most historians attribute the growth of Harvard's astronomy curriculum entirely to him. However, at the undergraduate level, the growth started a quarter of a century earlier in a small wooden building, close to Harvard Yard, called the Astronomical Laboratory. It was there that Robert Wheeler Willson instituted such a successful undergraduate program at Harvard that astronomy teachers at other colleges around the world imitated his innovative methods by using his laboratory apparatus and textbook.

Although Willson served as a tutor in physics briefly in the 1870s and an instructor in the 1890s, he probably would not have had much influence on the teaching of astronomy at Harvard if he had not secured a building for the Astronomical Laboratory in 1902 and received a professorship there in 1903. The brilliance with which Willson served in that capacity until his retirement in 1919 more than offset his rather undistinguished career in astronomy and physics in the nineteenth century.

In the first decade after his graduation from Harvard in 1873, his early interest in observing led him to three short-term astronomical positions: first at the Cordoba Observatory, then at Harvard College Observatory, and finally at Yale Observatory. His remarks about those positions in the first report for the Class of 1873 reveal that he did not at that time find in the stars what he was looking for:

"I spent the first year after graduation in the National Observatory at Cordoba, Argentine Republic, S.A., leaving home two days after Class Day, and arriving home again Sept. 20, 1874, to become an assistant at the

Cambridge Observatory ... [I took] charge of observations with the large equatorial, where I remained until the summer of 1875. Up to the spring of that year, I had been undecided whether I should make astronomy my profession; and, if Prof. Winlock had lived I daresay I should have remained at the Observatory. His death, however, removing a strong motive, I gave it up, and accepted the position of tutor in physics, which I now hold, and in which I expect to continue until further notice." (Ware, 1876)

Perhaps Arthur Searle inadvently discouraged Willson from continuing at Harvard College Observatory. After Joseph Winlock died, Searle became acting director until Edward Pickering took over in 1877. According to Bailey (1931), Searle was "one of the most genial and lovable of men," but he had "an appearance almost of gruffness." In the Harvard Archives there are three letters from Searle to Willson, dating from the summer of 1875 to the spring of 1876, about cleaning up his work. To a young man mourning the loss of his mentor, the formal tone of the letters may have seemed stern and unfriendly.

In the spring of 1878, after Pickering had been in office a year, Willson apparently sent him a notice of a new physics course in which he wanted to teach the students how to use the telescopes at the Observatory. Pickering replied:

"The clause relating to the Observatory in your announcement will be more definite if it is changed to 'with permission to use the smaller instruments of the Observatory.' To say 'the resources of the Observatory' might be taken for a promise of the unrestricted use of the Equatorial.

"I take the opportunity of assuring you that it will give me pleasure to provide you with any means at my disposal for carrying out your plan satisfactorily, and that I consider it desirable for the Observatory to connect itself with any astronomical instruction given to advanced students in the University."

It is interesting that initially Pickering seemed open to a liaison between the Observatory and the University. Before he came to Harvard, he had taught physics at Massachusetts Institute of Technology where he had developed some fine courses and a model students' laboratory. However, at Harvard he soon narrowed his goals for the Observatory to fundraising, data gathering, and analysis.

Meanwhile, in subsequent class reports Willson related his abiding interest in astronomy. In 1878 he wrote that he had gone to Texas to observe the total eclipse on July 29. And in 1883 he reported that he was "in New Haven at Yale Observatory since August 1881." While in New Haven, Willson married Annie Downing West. And in the spring of 1884, the Willsons went to Germany, where he began work on a doctorate in physics and chemistry. Having published his dissertation (Willson, 1886), he returned to Cambridge, apparently without any job offers. In 1888 he wrote to his class secretary:

"Since that time I have been in Cambridge at work on private scientific work, mainly in Electrical subjects. The authorities of the Jefferson and Physical laboratory now kindly allow me the use of a room and other facilities for carrying on my work begun in Germany."

Although Willson began to publish papers on his research both independently (1888; 1890a,b) and jointly with Benjamin Osgood Peirce (1889a,b), he sounded unenthusiastic in his report for 1891. He noted that the most exciting event in his life at that time was a serene fishing trip in Canada. And he related that he was still at Jefferson Physical Laboratory "where I have a room in which I occupy myself with various scientific researches of no very great interest to the general public."

Perhaps Willson's discontent with pure research led him to investigate the possibilities of again teaching at Harvard. At any rate, in 1891 he was appointed an instructor in physics and astronomy and began to build up a whole department of instruction in descriptive astronomy. In a posthumous sketch of Willson's life, his vigor in this endeavor was characterized as follows: "With remarkable ingenuity, originality, and untiring industry, he devised apparatus and methods that have proved their efficiency and have given to instruction in astronomy a marked impulse." (Dowse, 1923). By 1898 Willson was appointed assistant professor in astronomy and mathematics and by 1903, full professor of astronomy. His enthusiasm restored, Willson reported to his class in 1905 that he was "prepared to turn out competent astronomers in the shortest possible time, starting from the very depths of ignorance."

In addition to his own money, Willson secured funds for equipment and instruments for the Astronomical Laboratory from such generous and supportive benefactors as George R. Agassiz. In the Harvard Archives are some letters that Agassiz penned to Willson in 1907. In one he wrote "I enclose a check for $500. I trust that you will be able to get just what you want. Wishing you a pleasant summer ... " In another, he encouraged Willson to "drop me a line anytime that you want the money for that spectroscope, and I will send you a check." And at the end of the year Agassiz communicated " ... I want to hear how the telescope is working, and see if there is anything I can give it for a Xmas present." (Willson, 1877–1916)

As successful as he was in running the Astronomical Laboratory independent of the Observatory, Willson apparently still wanted a liaison between the two astronomical institutions and mentioned it to Solon Bailey's brother in 1911. In reply Bailey wrote the following letter:

"My brother mentioned to me your conversation with him in regard to some relation in teaching between the <u>Observatory</u> and the <u>Laboratory</u>. I must say that

I am entirely in sympathy with some such plan, but whether it could be arranged at the present time to advantage, I am not sure. I should be glad to talk it over with you at your convenience and get your views in regard to it. It may be that a plan could be arranged which would appeal to Professor Pickering."

If any plan was ever drawn up, it clearly was never implemented. Willson continued to teach astronomy independent of the Observatory until he retired in 1919, the same year that Pickering died. Through the establishment of the Robert Wheeler Willson Professorship in 1928, Annie Willson tried to perpetuate astronomy "at the College itself as distinguished from the Astronomical Observatory." (President, 1948a) Ironically, neither the undergraduate curriculum in astronomy nor the professorship continued in Harvard Yard. Ultimately, they both came under the direction of Harlow Shapley, Pickering's successor at the Observatory. Clearly, Harlan T. Stetson, Willson's successor in the Laboratory, thought that he should receive the named professorship (Stetson, 1928). However, for whatever reasons, he never did. In fact, for nearly 20 years no appointment was made until Bart J. Bok at Harvard College Observatory was named the first Robert Wheeler Willson Professor of Applied Astronomy in 1947. By that time, of course, the Astronomical Laboratory had been disbanded.

Besides developing the Laboratory and the first full-scale undergraduate curriculum in astronomy at Harvard, Willson pursued other interests such as collecting rare astronomical books. For an exhibition of such volumes at Harvard, Owen Gingerich wrote:

" ... the backbone of Harvard's collection of rare astronomical books was formed only in 1927, by the gift of 127 volumes from the library of Robert Wheeler Willson ... He prized especially highly a presentation copy of Galileo's *Dialogues*, published in 1632, inscribed presumably in the handwriting of its author.

"Unfortunately, no list was kept of the 127 books given in 1927 by Willson's widow, but his bookplate clearly identifies the volumes. The books themselves stand in mute testimony to a remarkably knowledgeable and thorough collector. Willson bought not only the well-known classics, such as those by Galileo and Kepler, but also the lesser and often rarer works required for a research collection. Examples of the latter are the earliest works on sunspots, published anonymously by Christopher Scheiner (the exhibition showed only the *De Maculis Solaribus* of 1612 but the collection also includes the slightly earlier, rarer, but less showy *Tres Epistolæ de Maculis Solaribus*).

"Willson copies were shown in nine of the ten major display cases of the 'Rara Astronmica'; they ranged

from a 1478 *Sphere of Sacrobosco* to Thomas Wright's
Original Theory or New Hypothesis of the Universe
(1750). The six titles exhibited that had gone up in
smoke in 1764 were all from the Willson library. Alto-
gether, seventeen Willson books made up a quarter of
the exhibition." (Gingerich, 1971)

As Gingerich noted rather cryptically, Willson's library replaced many early astronomy volumes that had been lost in the fire of 1764 at Harvard.

With his keen interest in navigation and surveying, it was but a small
step for Willson to transfer his skills from the sea to the air. Therefore, he actively
participated in the new developments in aviation. He not only took charge of
measuring the time and altitude of flights at early aviation meets (Willson, 1911),
but he also succeeded in developing a bubble attachment for the marine sextant
(Rogers, 1971). With that "bubble sextant" aviators could always determine the
true horizontal plane and therefore navigate even when clouds or darkness obscured
the earth below. The increasing importance of air navigation during World War I
brought quite a bit of attention to the Willson's device. In 1918 it was tested in
numerous flights at Langley Field. Basing his remarks on its use in taking more than
a thousand sights on the sun, moon, and stars in nine different types of airplanes
and seaplanes, Henry Norris Russell reported that

“The bubble sextant appears to leave little to be desired as an instrument for aërial navigation. It is light,
portable, reasonably rugged, and gives a precision of
observation much surpassing the limit set by the small
residual irregularities in even the best piloting." (Russell, 1919)

Smithsonian has at least four bubble sextants with the Willson attachment that
were manufactured by Brandis & Sons, Inc. of Brooklyn, New York as early as
1921. (Rogers, 1971)

Finally, Willson's interest in other languages and cultures led him
to investigate the mysteries of Mayan astronomy. The Harvard Archives has two
binders full of his calculations of eclipses followed by Venus conjunctions for the
period of the Mayan civilization. He also studied planetary tables, eclipses, and
the Dresden Codex. (Willson, 1924) Willson worked on this project steadily until
a short time before he died in November 1922. Harlan Stetson pulled together the
final manuscript and submitted it for publication in 1924.

This sketch of Willson has focussed mainly on the tangible aspects
of his life and legacy. But what made the undergraduate courses in astronomy at
Harvard so popular in Willson's day was the intangible aspect of his charismatic
personality. Throughout the reports for the Class of 1873, Willson was mentioned
as having spoken or sung at his class reunions. The Fiftieth Anniversary Report
summed it up best: "His histrionic and melodic gifts and his ever merry spirit and
generous heart are a grateful memory with all who really knew him." (Dowse, 1923)

Reference list.

Bailey, S.I. (1931). *The History and Work of Harvard Observatory 1839 to 1927.* New York, NY: McGraw-Hill Book Co., Inc.

Dowse, W.B.H. (1923). *Harvard College/ Class of 1873/ Report X/ Fiftieth Anniversary Report*, pp. 116–120. Privately Printed for the Class.

Editors, The (1943). Robert Wheeler Willson. In *Who was Who in America*, **1**, *1897–1942*, p. 1358. Chicago, IL: The A.N. Marquis Company.

Gingerich, O. (1971). Rara Astronomica. *Harvard Library Bulletin* **19**, 117–139.

Peirce, B.O. & Willson, R.W. (1889a). Note on the Measurement of the Internal Resistance of Batteries. *The American Journal of Science* **138**, 465–468.

Peirce, B.O. & Willson, R.W. (1889b). On the Charging of Condensers by Galvanic Batteries. *Proceedings of the American Academy of Arts and Sciences* **24**, 146–163.

Peirce, B.O. & Willson, R.W. (1895). The Temperature Variation of the Thermal Conductivities of Marble and Slate. *The American Journal of Science* **150**, 435–441.

Peirce, B.O. & Willson, R.W. (1898). On the Thermal Conductivities of Certain Poor Conductors, I. *Proceedings of the American Academy of Arts and Sciences* **34**, 1–56.

Pickering, E.C. (1877–1887). Letter books containing copies of manuscript outgoing correspondence from Harvard College Observatory. Harvard Archives, Pusey Library, Harvard University, Cambridge, MA.

President and Fellows of Harvard University (1948a). Robert Wheeler Willson Professorship of Applied Astronomy (1942). In *Endowment Funds of Harvard University*, p. 83. Cambridge, MA: Harvard University Printing Office.

President and Fellows of Harvard University (1948b). Robert Wheeler Willson Scholarship Fund in Applied Astronomy (1928). In *Endowment Funds of Harvard University*, p. 83. Cambridge, MA: Harvard University Printing Office.

Rogers, F.M. (1971). The Search for a Suitable Bubble Sextant/ Robert Wheeler Willson. In *Precision Astrolabe*, pp. 172–181. Lisbon.

Russell, H.N. (1919). On the Navigation of Airplanes. *Publications of the Astronomical Society of the Pacific* **31**, 129–149.

Searle, A. (1875–1876). Letter books containing copies of manuscript outgoing correspondence from Harvard College Observatory. Harvard Archives, Pusey Library, Harvard University, Cambridge, MA.

Stetson, H.T. (1918). The Astronomical Laboratory in War Time. *Harvard Alumni Bulletin*, **20**, no. 32, 620–624.

Stetson, H.T. (1923). Robert Wheeler Willson. *Popular Astronomy* **31**, 308–313.

Stetson, H.T. (1928). Twenty-five Years of the Students' Astronomical Laboratory, 1903–1928. *Harvard Alumni Bulletin*, **30**, no. 20, 588–594.

Stetson, H.T. (1930). The Teaching of Astronomy and the Astronomical Laboratory. In *The Development of Harvard University since the Inauguration of President Eliot*, 1869–1921, ed. S.E. Morison, pp. 303–306. Cambridge, MA: Harvard University Press.

Ware, A.R. (1876). *The First Triennial Report of the Secretary of the Class of 1873 of Harvard College*, pp. 25–26. Boston, MA: Franklin Press.

Willson, R.W. (1877–1916). Folders of miscellaneous manuscript materials such as observations, papers, and correspondence. Harvard Archives, Pusey Library, Harvard University, Cambridge, MA.

Willson, R.W. (1885). Ein empfindliches Galvanometer mit messbarem Reductions-factor. *Annalen der Physik und Chemie*, New Series **26**, 44–55.

Willson, R.W. (1886). Dissertation: Ueber ein empfindliches Galvanometer mit messbarem Reductions-Factor. Würzburg: Druck der Thein'schen Druckerei (Stürtz).

Willson, R.W. (1888). Mode of Reading Mirror Galvanometers, etc. *The American Journal of Science* **136**, 50–52.

Willson, R.W. (1890a). The Magnetic Field in the Jefferson Physical Laboratory. *The American Journal of Science* **139**, 87–93.

Willson, R.W. (1890b). The Magnetic Field in the Jefferson Physical Laboratory, Part II. *The American Journal of Science* **139**, 456–70.

Willson, R.W. & Peirce, B.O. (1897). Table of the First Forty Roots of the Bessel Equation $J_0(x) = 0$ with the Corresponding Values of $J_1(x)$. *Bulletin of the American Mathematical Society*, Second Series **3**, no. 4.

Willson, R.W. (1901). *Laboratory Astronomy*. Boston, MA: Ginn & Co.

Willson, R.W. (1905). *Laboratory Astronomy*, 2nd ed. Boston, MA: Ginn & Co.

Willson, R.W. (1908). *Times of Sunrise and Sunset in the United States*. Cambridge, MA: Harvard Cooperative Society.

Willson, R.W. (1910). The Natural History and Scientific Book Circular No. 145. *Astrophysical Journal* **32**, 326.

Willson, R.W. (1911). Determination of the Altitude of Aeroplanes. *Proceedings of the American Academy of Arts and Sciences* **47**, no. 2.

Willson, R.W. (1924). Astronomical Notes on the Maya Codices. *Papers of the Peabody Museum of American Archaeology and Ethnology, Harvard University*, **6**, no. 3.

THE GREAT MNEMONICS CONTEST

Owen Gingerich
Harvard-Smithsonian Center for Astrophysics
Cambridge, MA 02138,
USA

I first met Bill Liller when I came to HCO to work as an
assistant for Harlow Shapley in the summer of 1949. Bill had just
graduated, but was taking some math courses before going on to graduate
school at Michigan. Our paths intersected at various times, but
especially after 1960 when he returned to Cambridge as Professor;
he was on the examining committee for my doctoral thesis in the summer
of 1961, and, as Department Chairman, it was he who invited me in 1963
(at comparatively short notice) to teach the department's offering in
Harvard's General Education program, Natural Sciences 9.

I changed the name of the course to "The Astronomical
Perspective," and have been teaching it more or less ever since. I say
"more or less" because Bill actually taught three semesters of it.
And thereby hangs the tale about the Great Mnemonics Contest.

From time immemorial, or at least since Henry Norris Russell
supposedly invented the mnemonic early this century, students have
remembered the temperature order of the spectral sequence, OBAFGKMRNS,
by the phrase
Oh Be A Fine Girl, Kiss Me Right Now Smack!

By the late sixties, 1968 to be precise, when Bill was teaching the
spring term of Nat Sci 9, the standard mnemonic was considered too
sexist, so he offered a prize for a substitute. Not only did he
initiate the contest, but he suggested the mnemonic might be updated
by including Wolf-Rayet or W-type stars at the beginning. No one now
remembers which of the 19 entries won, and none seems wildly memorable,
but perhaps the prize (a Mars bar) went to
When Obstreperous Beasts Approach, Fragrant Geraniums
Knowingly May Receive Night's Stigmata.

In any event, the contest seemed like a good idea, and it has
become a permanent feature in the course. The second year produced what
remains one of the most memorable entries, and a haiku at that:
Oh Bring Another Fully/Grown Kangaroo,
My Recipe Needs Some./

College life quickly played a role in the entries, with
Oh Brutal and Fearsome Gorilla,
Kill My Roommate Next Saturday

and
>   Obese But Alluring Fertility Goddess,
>   Keep My Radcliffe Nymph Sensuous.

(A faculty member from another department suggested the last word might more appropriately be "Sterile.")

Politics also provided grist for the mill, with the following 1972 entry:

>   On Bad Afternoons, Fermented Grapes Keep
>   Mrs. Richard Nixon Smiling.

When the Massachusetts governor raised the drinking age in 1979, this one took the prize:

>   Ornery Bostonians Angrily Fight Governor King's
>   Measures Regarding Needless Sobriety.

And in the twelfth annual contest, the political relevance winner was

>   Own Bibles! Attack Foreign Governments! Kill Murderers!
>   (Reagan's Next Speech).

The all-time winners (so far) for astronomical relevance and memorability seem to be

>   Out Beyond Andromeda,
>   Fiery Gases Kindle Many Red New Stars

and

>   Only Bright Astral Fires Going Kaput
>   Make Real Neutron Stars.

In the sixteenth annual contest in 1985, a pair of enterprising classicists submitted the following entry, but we had to construct the mnemonic paraphrase:

>   Observate! Bellis Astris, Ferox Gradiuus
>   Kalendarum Medio Residae Noctis Scandet

or

>   Oh Beautiful Ascendant, Fiery Gladiator,
>   Knavish Mars Rises Nocturnally Skyward.

Two rhyming mnemonics have captured my fancy:

>   Oh Backward Astronomer, Forget Geocentricity!
>   Kepler's Motions Reveal Nature's Simplicity.

and rather more memorable,

>   Organs Blaring and Fugues Galore,
>   Kepler's Music Reads Nature's Score.

Not knowing that Bill Liller had inaugurated the contest with astronomical candy bars as prizes, we have traditionally awarded 20 or 30 all-different IBM cards, with the admonition that such ephemera will surely be great collector's items if saved long enough. In addition, we almost always have a booby prize to award--punches from IBM cards (which

are now extremely hard to find!)--for those that use my name, such as
            Owen (Boring, Aging, Fuddy-duddy) Gingerich
            Knows Most Regarding Natural Sciences.

So the contest goes on, and there seems to be no thought of dropping it,
especially when we get such novelties as
            Surprising New Reverse Mnemonics
            Keep Gingerich From Accepting Boring Options.

Not only does the annual competition continue here, but the publication
of selected winners in Physics Today, in Mercury, and in some
astronomy texts has insured that the contest is spreading beyond the
ivied walls of Harvard. Will Yervant Terzian challenge us to an
intercollegiate contest? Will Bill Liller send entries in Spanish?
Keep in touch!

(Photo by Matty Liller)

Bill Liller on Easter Island with his "little Schmidt" used for his observations of Comet Halley as part of the NASA International Halley Watch.

HETU'U RAPANUI: THE ARCHAEOASTRONOMY OF EASTER ISLAND

W. Liller
Instituto Isaac Newton, Ministerio de Educacion,
Casilla 8-9, Correo 9, Santiago, Chile

1 *PREAMBLE*
        Please forgive my audacity in presenting a paper at my own
festschrift, but I very much want to share with you the fascination and
the excitement of my latest field of study. Moreover, it is fitting, I
feel, to subject you, students of years sometimes long past, esteemed
colleagues, and good friends all, to the first verbal presentation of my
findings and conclusions. Do not spare me your searching questions and
critical comments: I have learned from you in the past; teach me more.

2 *AN INTRODUCTION TO EASTER ISLAND*
        The extreme isolation of Easter Island has made it a fasci-
nating socio-anthropological test-tube: for the approximately 1400 years
before Western man arrived in 1722, it was almost totally left alone by
the world to develop in its own special way. Little Pitcairn Island 1200
miles away (pop. 58) is the nearest inhabited land, and the American con-
tinent is nearly twice as far away. With only a very small dissent (led
by the distinguished world traveler, Thor Heyerdahl), the general consen-
sus among archaeologists and anthropologists is that Polynesians came
from the west to populate the island. Perhaps they were searching for
new islands; perhaps they were lost or blown off course. However, it is
well known that over a thousand years before Columbus first struggled
across the Atlantic, Polynesian navigators were criss-crossing the Pacific in double canoes that sometimes were large enough to carry several
families and a small contingent of chickens, rats and other tasty,
nutrious edibles (Makemson 1941, Bateson 1959, Lewis 1974).

According to legend, the first permanent settlement was established by
the *ariki* (chief) Hotu Matu'a who landed with his family at Anakena, one
of the few sandy beaches on the coast. The island possessed no atolls or
lagoons teeming with fish, but because of the sub-tropical temperatures
(aver. 22° C) and the moderate rainfall (110 cm y⁻¹), agriculture flourished and the population grew. On rare occasions other boats must have
arrived from the west, but the culture brought by Hotu Matu'a appears to
have totally dominated the island. The existence of other islands was
soon so completely forgotten that Easter Island never really received a
name. It was simply called "The Land" by the natives, and more recently
"Big Island" (*Rapa Nui*) to distinguish it from a smaller island (Rapa) to
the west. However, it can hardly be considered "big" in the usual sense;
its longest dimension is only 24 kilometers. Its size is roughly that of
Nantucket, the island where the Maria Mitchell Observatory is situated.

The colossal island statues -- *moai* in the ancient Polynesian dialect

spoken by the rapanui -- were carved out of volcanic tuff as monuments
to deceased *ariki*. Many were placed at the middle of sometimes very
long (>100 meters) and very elaborate platforms (*ahu*). Ahu Tongariki,
destroyed in 1960 by a tidal wave, once supported an estimated 20 moai;
more often only one or two statues adorned an ahu (Englert 1948).

As far as is known today, the island society quickly developed into the
classic pattern of workers and chiefs with the latter convincing (or
forcing) the former to move the multi-ton moai long distances and to
erect them on the ahu around the coast wherever settlements grew up.
Over 90 per cent of the ahu are located within a few meters of the shore
(see Fig. 1). The moai, when present, had their backs to the sea.

Carbon dating reveals that the earliest ahu were built no later than A.D.
800, but then in the 16th or 17th century, the island society abruptly
collapsed. Suddenly, the carving of statues halted, and subsequent
intra-island warfare reduced the rapanui people to the status of savages,
living fearfully in caves, rather than in carefully made thatched houses
by the magnificent ahu. Instead of ceremonial human sacrifices, there
was cannibalism. Reports from the first European visitors noted that
many statues were seen standing, but it seems that by 1840 all had been
toppled and many brutally mutilated (Van Tilburg 1986).

Perhaps more to Easter Island than to any other island in the Pacific,
Western man brought unspeakable misery. Beginning early in the 19th
century, ships raided the islands for slaves. The Americans came first,
but the worst came around 1860 when Peruvian ships carried away hundreds
of men, women and children. The strongest were sent to mine *guano* --
bird droppings -- on islands off the coast of Peru. Public outcry
reversed the process, but the few survivors who returned to Easter Island
brought with them some of the worst diseases of the civilized world.

Figure 1. - Easter Island, or Rapa Nui

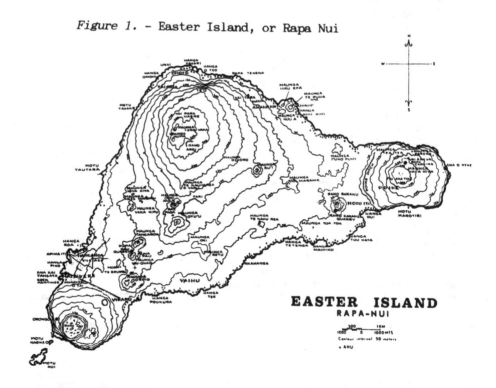

EASTER ISLAND
RAPA-NUI

Whereas once the island population was probably over 10,000, by 1877,
less than 200 remained. Consequently, the ethnology of Easter Island is
sparse; the elaborate hieroglyphic writing that developed in island
isolation is no longer understood (nor has it been deciphered); and the
meanings of many place names have disappeared (Barthel 1978).

3 ARCHAEOASTRONOMY BEGINS

Several of the members of Heyerdahl's expedition to Easter
Island in 1955-56 noticed, as they tried to reconstruct the prehistory of
the island, that there was evidence of astronomical planning in some of the
early architecture (Heyerdahl & Ferdon 1961). At Ahu Tepeu, Carlyle
Smith found that the eight or nine moai mounted on two long and massive
ahu all faced within a few degrees of the azimuth where the sun first
appeared at the summer (December) solstice. William Mulloy noted that at
Vinapu, a pair of beautifully constructed ahu side by side and both over
50 meters long were oriented so that to within one and three degrees,
respectively, perpendiculars to their facades pointed at the rising
points of the summer solstice and the equinoctial sun. And perhaps most
interestingly, Edwin Ferdon, Jr. discovered pecked into bedrock beside an
ancient ahu a cluster of four depressions, at least three of which were
clearly man-made. According to his measurements, three pairs of these
cavities were aligned with the solstitial sunrise points and with the
equinoctial sunrise direction. His conclusion was unequivocal: "From
direct observation, then, it can be definitely stated that the complex of
four holes...constituted a sun observation device". He referred to
them thereafter as the "sun stones" (p.228, Heyerdahl & Ferdon 1961).

These findings were not entirely surprising since the early Polynesians
were known to have been master navigators using sun and stars to find
their way through the maze of South Pacific islands. Moreover, on a sub-
tropical island where agriculture assumed vital importance, knowledge of
the advance of seasons would have been virtually essential.

Mulloy returned to the island a number of times afterward to carry out
further investigations and to search for other astronomical alignments.
His restoration work is well known, but more importantly, his archaeological studies were carried out with great care and intelligence. In
publications on the ahu Huri A Urenga, A Kivi, and Vai Teka, he reported
that these platforms were aligned solstitially or equinoctially, and that
at the first ahu, he had discovered another set of cavities resembling
the sun stones which he called a "solar ranging device". These, too,
were aligned with the crucial sunrise (and sunset) directions.

Following Mulloy's death in 1978, few additional archaeoastronomical
studies were undertaken until 1986 when the author arrived on Easter
Island as a part of a NASA project to observe Halley's Comet (Niedner &
Liller 1987). At that time Don Sergio Rapu Haoa, governor of Easter
Island and former graduate student of Mulloy's, discussed with me the
possibility that the magnificent ahu Nau Nau by the beach at Anakena
(which he had restored) might be oriented astronomically; and a few weeks
later, Dr. Georgia Lee of UCLA, the world expert on the island's petroglyphs, expressed to me her doubts about Ferdon's "sun stones" measurements, and suggested that I might check his results.

During the period my wife and I were on Easter Island (19 Feb - 3 May,

1986), there was time not only to remeasure most of the alignments mentioned above but to begin a systematic survey of the orientations of all known ahu. In an inventory made by Englert (1948), 243 are listed, but the more recent, partially completed Atlas Arqueologico de Isla de Pascua (Cristino *et al.* 1981) indicates that the true number may be over 500. We were able to measure about fifty ahu, mainly in the last two weeks when bright moonlight and bad weather made it difficult to observe Comet Halley. A further source of excellent data was several sets of aerial photographs made in the early 1960s by the Chilean Air Force. I am indebted to Governor Rapu for making them available to me. At a scale of approximately 1:17000, they show details to better than one meter.

The biggest stroke of luck was the revelation by Mulloy's daughter Brigid of the existence of a set of unpublished field notes made by her father in 1965 in which he recorded his measurements of the orientations of 226 ahu. (See Section 5.) The azimuths of the seaward facades or the long axes of the ahu were made with theodolite or alidade to the nearest arc minute and calibrated by direct observations of the rising or setting sun. The Mulloy family kindly made these notes available to Dr. Lee and me in late 1986, and she and I have recently submitted for publication a condensed and annotated version of his notes, including tabulation of all his measurements (Mulloy *et al.* 1987). Recently, Dr. Frank Morin, who has frequently assisted us in our work, has made the notes generally available by transcribing the entire set to PC-compatible diskette.

Consequently, my own program changed. In March 1987 my wife and I returned to Easter Island to examine in more detail those ahu that Mulloy found to be solstitially or equinoctially aligned, paying particular attention to prominent horizon features or other distant structures that could have been used as foresights.

Another interest of mine was to determine orientations of *hare paenga*, long, narrow houses with beautifully made stone foundations that are said to have been occupied by wealthy or highly ranked people. Some of these are over 50 meters long, and according to measures by Dr. Arne Skjolsvold of Heyerdahl's expedition, at least several are approximately aligned with the cardinal directions of the compass.

Finally, I have had the pleasure to work closely with Georgia Lee who has amassed an imposing collection of photographs and drawings of rapanui rock art. Included are numerous examples showing comet and crescent forms, starbursts, and rows and groupings of cupules (pecked cavities in stone) which may have been pre-historic calendars or drawings of constellations (see Figure 2). One site of interest was recorded by Katherine Routledge (1919): *papa ui hetu'u*, the rock (*papa*) for observing stars (*hetu'u*). She 'states, "About 200 yards from these boulders (direction unspecified) there is another engraved stone on which ten cup-shaped depressions are visible; this represented, it is said, 'a map of the stars'" (page 235). Lee and earlier Tom Hoskinson found an engraved stone about 200 yards from *papa ui hetu'u*, but the cupules thereon are in a row and can hardly be interpreted as a constellation. Another *papa* called Matariki, the Polynesian word for the Pleiades, has been described in detail to Lee by an islander and purportedly shows a dozen or so recognizable constellations (Taurus, Orion, Canis Major, Crux, etc.). However, attempts to locate it made both by Lee and by my wife and me have so far been unsuccessful.

4   *PROCEDURAL COMMENTS*

In this report, I will be concerned primarily with orientations and alignments of various sorts. Lee (1986) has presented a
thorough study of the rock art, and at the present time the two of us are
involved in a program of analysis of the astronomical representations.
We hope to report on this work in the near future.

I should remind the audience that at Harvard I was the Willson Professor
of *Applied* Astronomy (see paper by Barbara Welther in this volume), a
fitting title since several decades earlier, as a junior faculty member
at the University of Michigan, it fell to me to teach a course called
"Practical Astronomy", and in doing so, I became adept at the use of a
theodolite or alidade. Thus, measuring azimuths on Easter Island was a
straightforward matter, especially after the airport meteorologists had
put at my disposal a brand new Japanese-made theodolite.

Photographs also can be used to derive precise azimuths, and we now have
a collection of dozens of often spectacular 35mm slides taken just at the
moment of sunrise or sunset. These show the sun's limb clearly (usually)
with numerous horizon features which can later be used as standards of
reference. On one occasion the green flash was photographed, and the
following night a green flash was again seen and observed with an alidade. Similarly, the moon in the daytime and star trails in the dusk or
at night with moonlight can also be used to establish an astronomical
frame of reference. On three occasions rainbows (angular radius of
yellow light = 41°20' from the sub-solar point) have been photographed
from strategic locations to provide accurate orientation data.

For much archaeological work, a good magnetic compass gives sufficiently
accurate directional information, but the navigational charts of Easter

Figure 2. - Some possible astronomical petroglyphs (Lee
1986). The scale averages about 25:1.

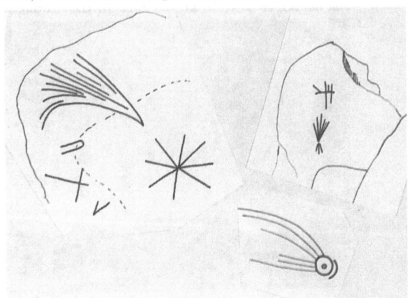

Island warn sailors that magnetic anomalies exist both on the island and near it. Therefore, we have repeated all of the earlier measurements even though they had been carefully made by experienced field workers.

In the discussion that follows, we will consider primarily the rising and setting points of the sun, corrected for (1) atmospheric refraction, (2) dip of the horizon produced by earth curvature, and (3) the slowly changing (0.13° per millenium) obliquity of the ecliptic. The assumed date will always be A.D.1000. The rising points of bright stars and the lunar standstills will be considered only briefly.

During the course of the year 1000, the azimuth of the point of sunrise swung back and forth between 63.29° at the winter (June) solstice to 116.71° at the summer (December) solstice. (Values correspond to the island center and +0.5° altitude. At this altitude the unrefracted sun would be 90° from the zenith.) The astronomer-priests of the island may have wanted to mark several dates on this celestial calendar; which ones we do not know. Numerous stone towers, some of them elaborately constructed, dot the island and many could have served as indicators of special dates, much the way important dates are indicated on modern calendars with red letters. However, knowing the directions to the extreme azimuths and the mid-point (90°) was probably also desirable to the islanders with their long heritage of navigational skills. It is with these directions, corrected for atmospheric refraction and the effects of local horizon altitudes, that we will be primarily concerned.

When considering the orientations of stone platforms, we will usually be fortunate to arrive at accuracies of better than a degree, but we will work to and hope for ±0.1°. Distant horizon features are more precisely measured, and here we will carry along an extra decimal (±0.01°).

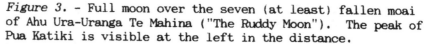

Figure 3. - Full moon over the seven (at least) fallen moai of Ahu Ura-Uranga Te Mahina ("The Ruddy Moon"). The peak of Pua Katiki is visible at the left in the distance.

In 1967 Alexander Thom published an important but controversial book on the archaeoastronomy of the British Isles, important because over a period of many years he had quietly and carefully measured and analyzed some 260 alignments at 145 sites where there were standing stones, henges, tumuli, and rings; and controversial because some felt that he had subconsciously biased his sample by selecting sight lines that gave him the desired results. Much good has come out of the subsequent, sometimes tense discussion (see, for example, Heggie 1982) because archaeoastronomers now take considerable pains to avoid selection effects and treat their results as objectively as possible. The author of this report has, at no time, made exception to these sound rules.

On many occasions an assistant was indispensable, and my wife Matty Pickhardt de Liller tirelessly stood ready to take photographs, time observations, hold ranging pole, or do whatever else was required. One of her special talents was finding badly destroyed ahu, barely distinguishable from the near-infinity of rocks that cover the island. In this report, the pronoun "we" is in no way used editorially.

5   THE MULLOY SURVEY OF AHU ORIENTATIONS

Not surprisingly, spot checks of Mulloy's unpublished 1965 field notes reveal that his measurements are of exceptionally high quality. He was a distinguished archaeologist who studied several ahu remains in great detail. With two rapanui assistants, he began at the first ahu listed in Englert (1948) and worked down this valuable inventory, which translates into traveling around the island clockwise beginning in the southwest corner. At each ahu well enough preserved to present a recognizable alignment, he measured with a surveyor's alidade the (usually) seaward facade or the long axis and referred all measurements to readings taken of the sun near the horizon.

The first conclusion that one draws from Mulloy's notes, apparent from even a casual inspection of the azimuths, is that a sizable majority of ahu, approximately 85 per cent, are aligned with their facades parallel to the neighboring shoreline. The two solstially aligned ahu at Tepeu studied by Smith are included; also the pair at Vinapu. But as Mulloy pointed out, the islanders may have selected these sites so that the ahu "could be so astronomically oriented and faced towards the sea at the same time" (p.94, Heyerdahl & Ferdon 1961). However, as Mulloy surely understood, there was no proof that this was indeed the case.

In searching for ahu that were *intentionally* oriented towards critical sunrise/sunset azimuths, we will first focus our attention on astronomically oriented ahu that (1) are not aligned parallel to the coast, (2) are located well inland, or (3) show some other evidence for a conscious astronomical intent by the builders. In this last category we might include, for example, ahu whose names carry astronomical connotations, or which are located such that directions towards conspicuous horizon features coincide with special solar azimuths.

In Table 1 are listed all the solar-oriented ahu from Mulloy's survey that fulfill one the first two conditions. The first column gives the number in Englert's listing (E) or in Mulloy's notes (M). Names in parentheses refer to the ahu location and may not have been of the ahu *per se*. The quantity A_{perp} is Mulloy's listed azimuth of a *perpendicular* to the feature measured (a facade, the edge of the central platform that supported the moai, a long central axis).

Mulloy generally listed directions of perpendiculars reasoning, no doubt, that if solar orientations were intended, then the moai would have faced the corresponding sunrise or sunset. The previously mentioned ahu at Vinapu do not strictly fit this category; their moai had their backs to the rising sun. Instead, they faced the smooth slopes of a nearby mountain, Rano Kau behind which the setting sun disappeared at an elevated altitude of more than 7°. Perhaps there once were stone markers on the apparent horizon to indicate sunset position on important days of the year; if so, their presence is not known or recognized today.

Later we will consider the orthogonal possibility: that the ahu axes pointed in the critical directions. For now we note that some of the ahu of Table 1 were never intended to support moai. These include the *poepoe* that have been described as boat-shaped and are strongly symmetrical about their long axes. Many are directed perpendicularly towards the coast. For these Mulloy listed only the axial azimuths; for the sake of consistency we tabulate here the perpendicular directions.

The azimuthal criterion for inclusion of an ahu in the table was ±5° of one the six rising or setting solstitial or equinoctial azimuths, plus 0-180°, due north-south. The azimuths of the "megalithic equinox", coming when the sun is at declination +0.51° exactly a quarter year after summer solstice, was also considered as a legitimate significant direction. Ahu E144 has been included because to the west the horizon rises to an altitude of 2.0°, resulting in an equinoctial sunset that would have occurred at an azimuth of 270.87°.

Table 1. - Astronomically Oriented Ahu: List 1 (from Mulloy): Inland Location or Not Paralleling the Shore

No.	Name	A_{PERP}	Meas'd	L	Remarks
E5	Makere	244.2°	Facade	8m	Sunrise at 2.5° altitude.
M245		269.7	W. facade	6	North end Hanga Pico Bay.
E6	Te Ata Hero	292.1	W. facade	31	One moai. Poor condition.
M252	A Kivi	273.2	E. facade	25	7 moai restored by Mulloy & Figueroa (1961, 1978).
E51	A Tanga	89.7	Axis	26	Poepoe in exc. condition.
E60		116.5	Facade	12	May be a *hare moa*.
E61		90.6, 88.1	Platform, N. facade	50	Complicated form; possibly two ahu.
E98	(Hue Mariki)	120.6	Axis	20.5	Poepoe, exc. condition.
E106	Hekii 1	68.5, 66.3	Old facade, New facade	88	Massive construction, 5m high. At least 6 moai.
E107	Hekii 2	88.7, 76.0	Platform, E. Facade	35	20m in front of facade of Hekii 1. Four moai.
E124	(Te Pupuhi)	88.6	Facade	33	Much destroyed. May have been a poepoe.
E144	(Paravo)	95.3	Axis	16	Poepoe. 1 km from coast.
E150	Huri Avai	116.5	E. facade	8	Semi-pyramidal form.
M265	Ruru O Ao	89.8	E. facade	32	100m N of E188. One moai.
M270		182.8	Facade	14	Crude construction.
E230	Mata Ketu	64.9	Facade	6	1.1 km inland. Four moai.

6 *RESULTS AND DISCUSSION*

Given a set of randomly oriented ahu, we would expect 22.2% to fall in these eight ±5° bins. Approximately 35 of the ahu measured by Mulloy either have their long axes at an angle of more than 20° to the adjacent shoreline, or lie more than a kilometer from the coast. Thus, on a statistical basis we can expect that about eight of the ahu of Table 1 are intentionally oriented with respect to the crucial solar directions. It remains to search for information to enable us to select out those ahu that were intentionally oriented astronomically.

To try to determine if some ahu had been carefully located so that prominent horizon features also were aligned with the significant directions, we visited each of the ahu in Table 1 and measured directions of distant mountains, distinctive notches between hills, edges of cliffs, etc. Although the reductions have not been fully completed, we can report that so far there are no obvious foresights for any of Mulloy's ahu. (Later, we will describe two ahu, not included in Mulloy's field notes, that do have natural "pointers".)

The ahu that Smith called Hekii 2 (E107) has a curious construction feature that is known to exist in only one other ahu, Huri A Urenga. As we will describe later, this latter ahu (not in Mulloy's field notes) almost certainly was intentionally oriented solstitially. Whereas the overall orientation of Hekii 2 is decidedly not solar (76.0° azimuth), the well-constructed central platform is skewed 12.7° to this direction so that its moai would have faced azimuth 88.7°, just 1.0° to the left of the rising megalithic equinox. Inspection of the site reveals no terrain feature that would make such an unusual design necessary. Moreover, the larger companion ahu, Hekii 1 (E106), is itself solstitially oriented, the re-faced (western) section of the seaward facade most accurately so.

From Hekii 1 one sees an ocean horizon to the north through the east until azimuth 93°; then the northern cliff of the Poike Peninsula (see Fig. 1) rises abruptly, and the apparent horizon climbs dramatically to

Figure 4. - Sunrise, March 5, 1987, as seen from the ahu Hekii 2. To the right of the sun protrude three cinder cones on the slopes of Pua Katiki on the Poike Peninsula.

an altitude of 3.6° at azimuth 112.5°. Summer solstice sunrise would
have occurred 2.4° to the right of this point well marked by a small
tree-filled crater known as Pua Katiki. (Later, we will describe some of
the frightening antics of the god known as Katiki.) Except for this one
approximate alignment, no prominent horizon features mark the critical
sunrise directions. But as Figure 4 shows, there are several eminently
conspicuous landmarks which could have been used to mark clearly those
dates of ceremonial importance and to gauge the year's progress.

In his book on the history and ethnology of the island, Englert (1948)
lists as the three "most outstanding monuments" Hekii, Vinapu, and
Tongariki, and in these three monuments, there are a total of five
platforms. We have already noted Vinapu's astronomical orientations, and
according to Cristino et al. (1981), as well as our own theodolite
measurements, a perpendicular to the seaward facade of Tongariki had an
azimuth of 120.0°. Thus, we have the intriguing result that the five
platforms of the three most outstanding monuments of the island are all
solstitially or equinoctially oriented. The probability of this
happening accidentally for any five randomly selected ahu is 0.0005.
Could it be that many of the larger and better built ahu are
astronomically oriented?

Since none of the remaining ahu in Table 1 are particularly outstanding,
we turn to Englert's second rank list: 12 ahu that "do not stand out so
much for their perfection and dimensions as the 3 mentioned, but that
also have considerable value." All were measured by Mulloy, and only one
has a facade perpendicular oriented within 5° of a significant solar
direction, the afore-mentioned Ahu Tepeu. (But see, also, the discussion
below on Ahu Ura-Uranga Te Mahina.) The two platforms of Tepeu are
oriented with facades closely parallel to the coastal cliff on the northwest shore of the island. However, the general alignment of this coast
averages an azimuth of about 25°, meaning that most of the nearby ahu
would be expected to have their perpendiculars oriented towards the
rising December solstice. Indeed, this is exactly what is found from
consideration of Mulloy's measurements.

What other hints do we have? Many ahu never were intended to have moai,
but we can readily imagine that the image of a moai facing towards the
rising or setting sun at a crucial time of year would have carried an
enormous aesthetic impact. We note that at least five ahu in Table 1
once had moai. It would seem to us that these are the best candidates
for ahu that were intentionally aligned with solar azimuths.

Several ahu have names that retain an astronomical flavor, and we paid
special attention to three ahu, not eligible for inclusion in Table 1,
that contain the words ra'a (sun) or mahina (moon). All three have
interesting orientations: (1) A line parallel to the facade of E227, Rua
Tau Ra'a ("Deep Red Sun") has an azimuth of 91.7°. (2) As seen from E115,
Ra'ai (perhaps a variant of the spelling of ra'a, according to a linguis-
tic expert on the island), the December solstice sun rose directly over
the crater Pua Katiki (see Figure 6) and set directly behind the crater
at the summit of Maunga Pui, the island's third tallest mountain. And
(3) one of the facades of E207, Ura-Uranga Te Mahina ("The Ruddy Moon")
points 6.2° to the right of the rising winter solstice and just 0.3° to
the right of one of the lunar standstills (see Section 9). These
intriguing coincidences suggest directions for future work.

Finally, we consider the possibility that the islanders constructed some ahu so that *parallels* to their long axes or facades were aligned with the critical solar directions, i.e., so that they could sight along the straight edge of the facade to determine the crucial azimuths. Table 2 lists those ahu that could have served this function, again to within ±5°, and that are either not aligned parallel (>20°) to the neighboring shore, or are located well inland. Two ahu, E76 and M261, have their axes roughly parallel to the nearest respective coasts, but they have been included owing to their moderately large distance from the shore.

A smaller number of ahu appear in Table 2 partly because we list only the solstitially aligned platforms. (Table 1 included ahu that faced *both* east-west and north-south.) Statistically, we should expect to find four ahu in Table 2, and if E76 and M261 are omitted, we indeed find exactly that number. Thus, there is no reason to believe that any ahu were intentionally oriented so that their long axes or long facades pointed in astronomically significant directions.

Table 2. - Astronomically Oriented Ahu: List 2 (from Mulloy): Inland Location or Not Paralleling the Shore

No.	Name	A_{PERP}	Meas'd	L	Remarks
E48	Rua Motu	331.9°	Axis	18m	Rectangular type.
E65	Motu Rau	150.7	Facade	30	Platform but no moai.
E76	(Vai Tara Kai Ua)	30.1	Facade	34	Poor condition. One moai. 700m inland.
M261	Te A Kava	30.7	N. facade	48	Two moai. 700m inland.
E188	Parai A Ure	206.5	Facade	30	Semi-pyramidal form.
E239	Motu Hitara	156.6	Facade	20	Platform but no moai.

Figure 5. - A solitary moai, recently re-erected, stands in front of the devastation that once was the great Ahu Tongariki which had perhaps as many as 20 moai. The sacred long-extinct volcano Pua Katiki rises in the background.

7   *INLAND AHU*

        During our several visits to Easter Island, we measured the
orientations of some of the inland ahu neither in Englert's inventory nor
in Mulloy's coastal survey. A few had been measured at other times by
Mulloy who no doubt wondered, as we did, if astronomical considerations
dictated the orientations and locations of these structures.

Well inland and nearly two kilometers north of the eastern end of the
recently enlarged airport runway stands a solitary moai (see Figure 7) on
an ahu studied and restored by Mulloy in 1969-70 (Mulloy 1975). Its
name, *Huri A Urenga*, means either "toppled by Urenga", Urenga being the
name of a person or group; or "pointing toward Urenga", where now Urenga
might be a place name, or conceivably a word for the direction of the
sunrise at the winter solstice. Mulloy's compass measurements, refined
optically by Liller & Duarte (1986), show that a perpendicular to the ahu
facade points at azimuth 64.8°, closely towards the point of the rising
sun at the winter solstice sun (64.0° over the ocean).

Given that there are hundreds of ahu on Easter Island, one normally would
not regard this single solstitial orientation as extraordinary. However,
Mulloy discovered a curiosity which alerted him to the likelihood that
the orientation was in fact intentional. Like Hekii 2, the central platform on which the moai was erected was conspicuously skewed (20°) to the
general plan of the stone-paved plaza in front of it. Mulloy's suspi-
cions increased when he discovered beside the ahu plaza a set of five
cavities pecked into bedrock that, according to his compass measurements,
indicated to within a few degrees the directions to the three critical
solar directions *plus* the north-south direction. He christened this new
group of man-made cupules a "solar ranging device".

Figure 6. - The view from Ahu Ra'ai showing where sunrise
would have occurred at the equinox and the summer solstice.

Figure 7. The moai of Ahu Huri A Urenga. Many lines of evidence now indicate that when it was erected it was intentionally oriented to look directly at the rising sun at the winter solstice. Its height is 3.34 meters; its estimated weight: 10 tons.

When the cavity alignments were remeasured accurately (Liller & Duarte 1986), the agreement was found to be considerably better than Mulloy had suspected: three of the four directions were indicated to accuracies better than a degree and the fourth to within 1.8°. Additionally, a *fifth* pair of depressions defined a direction that lay within 0.3° of parallelism to the plaza wall.

Other alignments at Ahu Huri A Urenga were discovered by Duarte and me: to within a few tenths of a degree, the summits of two prominent hilltops marked the directions of the rising winter solstice and the setting equinoctial sun as viewed from the moai. Additionally, to the east at an azimuth of 91.6° was found another ahu, also with one moai, and almost exactly due west (270.8° azimuth) lay a second ahu, again with a single moai. Thus, at least eight independent solar sightlines exist at Ahu Huri A Urenga -- plus the north-south direction.

These findings forced us to conclude that the ahu designers had consciously searched for a site from which distant hills would provide crucial sightlines, and then, upon locating it, they marked the location and orientation of the ahu and the significant directions by pecking cupules into the rock outcrop. Further details can be found in our report (Liller & Duarte 1986).

In a recent development, one of the island elders, Alberto Hotus, provided Georgia Lee with a detailed map showing the locations of ancient tribal boundaries said to have been established by Hotu Matu'a for his children. It is noteworthy that the location of Ahu Huri A Urenga had major significance: it marked the corner of four important territories in the island's southwest corner. It seems that this relatively small and humble-appearing ahu with its single moai achieved a lofty status in its day. We should note that one of the solstitially aligned inland ahu of Table 1, E230, Mata Ketu ("Uplifted Eyes"), is located 1400 meters from Huri A Urenga at an azimuth of 105°. It is probable that there existed a clear line-of-sight between these two similarly oriented ahu (trees now block the view), but whether there was any kind of astronomical collaboration between the two remains to be learned. Unrestored Mata Ketu awaits detailed study.

One of the most photographed ahus on the island is A Kivi whose seven re-erected moai (by Mulloy & Figueroa 1961, 1978) face toward the distant sea. See Figure 8. Mulloy and Figueroa noted two interesting orientations here: first, a perpendicular to the platform facade was directed 3.2° to the left of due east; and second, another ahu, Vai Teka with one moai, was (and is) clearly visible 770 meters to the west in a direction 5.9° to the left of due west. (The measurements made by Mulloy and Figueroa differ slightly from ours; the numbers quoted are based on our own optical measurements.) Their suggestion was that equinoctial orientations had been intended, noting that due east of A Kivi, the slopes of Maunga Terevaka created an elevated horizon at an altitude of 3.1° as seen from A Kivi, thereby shifting the apparent rising point to the left. However, the shift is too small to make up the difference; if there had been astronomical intentions, they were not very accurately carried out.

Very close to the exact center of Easter Island lies the remains of an ahu called *Moroki*, meaning "Well-constructed". According to the Atlas Arqueologico (Cristino *et al.* 1981), its orientation was approximately

solstitial, but dotted lines suggested uncertainty. The question of why
this most inland ahu was solstitially aligned (possibly) tweaked our
curiosity. On visiting the site, we discovered why dotted lines had been
used: the platform is almost totally buried beneath topsoil and tough
wild grass. Without permission from the Chilean park service, we could
do little more than yank up clods of grass. Still it was possible to
mark out roughly the alignment of the platform and measure an approximate
azimuth. The single moai, now lying face down on top of the mound, faced
3.9° to the right of the direction at which the sun rose over the ocean
at the winter solstice. To the east and west, none of the prominent
landmarks (including Pua Katiki) marked other significant directions.
Until an excavation is carried out, we must leave the conclusion hanging:
perhaps Moroki once was a solar observatory.

8 SUN STONES, BIRD MEN, AND SOLAR ECLIPSES

In reality, the first archaeoastronomical study that we
made, together with Drs. Lee and Morin, was to measure precisely the
directions indicated by Ferdon's "sun stones" at Orongo. This important
ancient ceremonial village is perched dramatically at the edge of a deep
1200-meter-wide crater at the summit of Rano Raraku, the long-extinct
volcano in the southwest corner of the island.

There are good ethnographical reasons to expect a calendric device to be
located at Orongo since it was from here that the annual "birdman" contest was supervised and observed. The winner of the contest was the
first one to bring back an egg laid by a sooty tern on the offshore islet

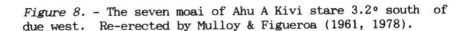

Figure 8. - The seven moai of Ahu A Kivi stare 3.2° south of
due west. Re-erected by Mulloy & Figueroa (1961, 1978).

of Motu Nui ("Big Isle"; see Fig.1). Clearly, one would have wanted to know accurately the time when the birds were expected to arrive and nest.

Our measurements of the sun stone directions strongly contradict Ferdon's conclusion that they "definitely...constituted a sun observation device". We found that the directions differed from the significant solar directions by 8, 12, and 16°. Only if there once had been an immense structure rising many meters above the level of the sun stones would they have worked in the manner suggested. Perhaps they were used to mark other dates on the calendars; perhaps they were used simply to support some sort of structure made of sticks and stakes. It has been reported that cadavers were sun-dried, hanging from such a framework. But never were the "sun stones" used as a solstitial and equinoctial device. Details of this work appear elsewhere (Lee & Liller 1987a,b).

During these investigations, an azimuthal point of reference was established in the direction of distant Pua Katiki that rises majestically out of the eastern peninsula (Poike) of Easter Island (see Fig. 1). The small crater at the summit was alleged to be the home of the evil god Katiki mentioned earlier. It seems that this god would, on occasion, eat the sun or the moon, sometimes partially, sometimes totally. Hence, Poike's Peak must have been regarded with fearsome reverence.

As seen from the sun stones, Katiki's crater home lies at azimuth 63.81°; sunrise at the winter (June) solstice in the year A.D.1000, occurred over Poike's Peak at an azimuth of 63.58°. As Figure 9 shows, the event must have been awesome for the ancient *rapanui*. Clearly, there was no need for a man-made device to mark this date. The sacred mountain must have been revered for its god-given ability to mark the shortest day of the year -- and to shelter the evil god Katiki.

There is excellent reason to believe that a special god once existed to account for eclipses: in the years A.D. 760-770, five total or annular solar eclipse paths passed either over Easter island or very close to it;

Figure 9. - A computer rendition of winter solstice sunrise over Pua Katiki as seen from the "sun stones" in A.D.1000.

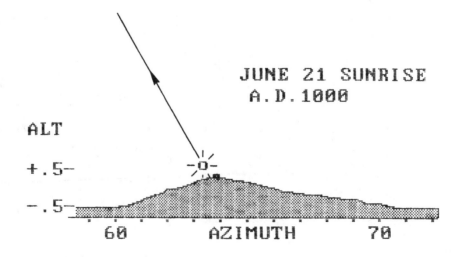

the most distant "near miss" was a scant 85 km away. Inspection of any
of the standard eclipse *canons* shows that such a profusion of eclipses is
extremely rare. In fact, no other similar grouping of eclipse paths can
be found at any other spot of the world during the first two millenia
A.D. (Liller 1986). The islanders must have felt extremely edgy by the
end of that unsettling decade. One can even imagine that an intense
activity of monument building began as a response to these happenings.

The next total solar eclipse came in January, 837 A.D., a year that surely was remembered by the islanders for yet another unsettling event: in
April Halley's Comet approached closer to the earth and became brighter
than ever before or since in its recorded history. According to Yeomans,
et al. (1986), Halley's reached an awesome magnitude -3.9 and passed
close to the Easter Island zenith at midnight with a tail length greater
than 90°. Those of the younger generation who might have been skeptical
of the skylore tales of their elders must have quickly become believers.

9 ODDS AND ENDS
Although we have not here considered alignments with the
rising and setting points of the moon and brighter stars, we will note
that one poepoe ahu may be named after a star, A Tanga (E 51). According
to Englert (1948), the ancient rapanui name for Antares was, possibly,
Rei A Tanga. ("Rei" means necklace.) Ahu A Tanga appears in Table 1
(A_{PERP} = 89.7°), but since the rising azimuth of Antares in A.D.1000 was
117.5°, it is difficult to see that a connection existed between ahu and
star. However, Antares must have been an important star: in A.D.1000 its
declination almost exactly equalled that of the sun at the December
solstice, and furthermore, it passed close to the zenith of Easter Island
(3.0° a millenium ago; 0.8° now). Perhaps there is a connection between
one (or both) of these facts and northward-pointing Ahu A Tanga.

The azimuths of the rising moon at its standstills were, in A.D.1000,
57.3, 69.2, 110.8 and 122.7°. None of these directions appear to be
over-represented in our compilation of Mulloy's data, but as we mentioned
earlier, the impressive ahu with a name containing the *rapanui* word for
moon, Ura-Uranga Te Mahina, has a facade that is aligned with azimuth
69.5°, a scant 0.3° away from a rising standstill.

The hare paenga, thatched houses with well-made curbstone-like foundations, are scattered all over the island. Their exaggerated elongated
shape -- sometimes several dozen meters long by 1.5 or 2 meters wide
-- interested us, and whenever we encountered one, we would measure the
the long axis azimuth. From a consideration of the 65 hare paega for
which we now have measurements (some from publications), two surprising
results have emerged: first, all seven that are longer than 20 meters
have perpendiculars to their axes lying within 5.8° of the significant
solar directions or north-south. Since the entrances to these long
houses were at the sides, one now should ascertain if, as one might
expect, the doors faced the sunrise, rather than the sunset.

Secondly, there is a high proportion of house lengths which are close
multiples of six meters, suggesting that there was, at least in the
construction of hare paenga, a standard of length. Such a concept has a
logical foundation since the foundation curbstones were built, on order,
at a distant quarry (McCoy 1976). Future investigators of Easter Island
should consider carefully the possible existence of a "rapanui yard".

10   CONCLUSIONS AND FUTURE WORK

Orientations of approximately 300 ahu have now been measured on Easter Island, the majority by Mulloy (Mulloy et al. 1987). From a systematic and unbiased consideration of these measurements, it would appear that there exist at least eight distinct sites with a total of 14 individual ahu that exhibit solar alignments supported by some kind of evidence that the alignments are more than happenstance. These ahu appear in Table 3 where we list each first by name, then by Englert, Mulloy, or Atlas number. The remaining columns give the number of ahu in each complex, whether the indicated directions are toward a Solstice or an Equinox, and finally the oriented structure or structures. The order is roughly geographical, not one of promise. Inclusion of Mata Ketu and Moroki is probably supported more by intuitive feelings than cold, hard facts -- or perhaps by ignorance since little is known about either site.

Ahu Huri A Urenga must now rank as one of the best-authenticated pre-historic observatories in the world. Its many solar sight lines and pointing devices, its use as an important land marker, and perhaps even its name all provide strong evidence that it was planned and built as an observing station. Its solstitially oriented moai appears in Figure 7.

We have just begun to survey orientations and lengths of *hare paenga*, the long thatched houses with well-constructed foundations. First indications suggest that solar orientations were intended for the longer houses, and their lengths seem to be quantized. Further work is planned.

11   ACKNOWLEDGEMENTS

Up near the top of my thank-you list must go the archaeolo-gists who first got me into this fascinating business: Georgia Lee and Don Sergio Rapu Haoa. The Mulloy family, especially Mrs. Emily Ross Mulloy and Mrs. Brigid Mulloy Robinson, have been more than generous and exceptionally patient with my numerous questions. By supporting me for other reasons, I must thank NASA, J.C. Brandt and M.B. Niedner, Jr. Gon-zalo Alcaino, Director of the Instituto Isaac Newton, has also provided solid support, both financial and moral. Many have encouraged me and given me good constructive criticism: John Carlson and Charles Peterson have been especially helpful. Haldan Cohn's critical reading of this paper has improved it substantially. We have numerous people on *Rapa Nui* to thank, particularly Julio Duarte, Urbano and Jacobo Hey, Lilian Gonzalez, and Javier Labra. Translations of place names were kindly provided by Cynthia Newson Rapu and Reina Rapu. But I save my deepest thanks for the end: to my wife Matty who was, simply, indispensable.

Table 3. - The Eight Most Promising Solstitial or Equinoctial Solar Sites

Name	Desig.	No. of Ahu	Dir.	Structure(s)
A Kivi	M252	2	E,E	Platform, 1 other ahu
Tepeu	E32,33	2	S,S	Both platforms
Hekii	E106,107	2	E,S	Both platforms
Tongariki	E158	1	S	Platform
Vinapu	E242,243	2	E,S	Both platforms
Mata Ketu	E230	1	S	Platform
Huri A Urenga	5-297	(3)	E,S	1 Platform, 2 nearby ahu
Moroki	10-305	1	S	Platform

12 *REFERENCES*

Barthel, T. (1978). The Eighth Land: The Polynesian Discovery and Settle-
      ment of Easter Island, translated by A. Martin. Honolulu:
      University Press of Hawaii.

Bateson, F. (1959). Publ. Astron. Soc. Pacific, 71, 187.

Cristino, C., Vargas, P., & Izaurieta, R. (1981). Atlas Arqueologico de
      Isla de Pascua. Corporacion Toesca, Santiago, Chile.

Englert, S. (1948). La Tierra de Hotu Matu'a. Historia, Etnologia y
      Lengua de Isla de Pascua. Imprenta y edit. "San Francisco"
      Padre las Casas, Chile.

Heggie, D. C. (1982). Archaeoastronomy in the Old World, edited by D.C.
      Heggie. Cambridge: Cambridge University Press.

Heyerdahl, T. & Ferdon, E. (1961). Archaeology of Easter Island: Reports
      of the Norwegian Archaeological Expedition to Easter Island
      and the East Pacific, Vol.1., School of American Research
      and Museum of New Mexico, Santa Fe.

Lee, G. (1986). Easter Island Rock Art: Ideological Symbols as Evidence
      of Socio-political Change. Ph.D. Dissertation, University of
      California, Los Angeles.

Lee, G. & Liller, W. (1987a). Easter Island's "Sun Stones": A Critique.
      Journal of the Polynesian Society, 96, 81.
      _____(1987b). Easter Island's "Sun Stones": A Re-evaluation.
      Archaeoastronomy, Supplement to the Journal for the History
      of Astronomy, No. 11, p. S1.

Lewis, D. (1974). Voyaging Stars: aspects of Polynesian and Micronesian
      Astronomy. Phil. Trans. Royal Society of London A., Vol. 276.

Liller, W. (1986). Celestial Happenings on Easter Island: A.D. 760 - 837.
      Archaeoastronomy, 9, 52.

Liller, W. & Duarte D. (1986). Easter Island's "Solar Ranging Device",
      Ahu Huri A Urenga, and Vicinity. Archaeoastronomy, 9, 52.

Makemson, Maude W. (1941). The Morning Star Rises. New Haven: Yale Uni-
      versity Press.

McCoy, Patrick C. (1976). Easter Island Settlement Patterns in the Late
      Prehistoric and Protohistoric Periods. Easter Island Commit-
      tee, Intl. Fund for Monuments Inc. Bulletin Five. New York.

Mulloy, W. (1975). A Solstice Oriented Ahu on Easter Island. Archaeology
      and Physical Anthropology in Oceania, 10, 1.

Mulloy, W. & Figueroa, G. (1961). Como fue restaurado el Ahu Akivi en La
      Isla de Pascua, Boletin No. 27, Universidad de Chile. Santiago.
      _____(1978). The A Kivi-Vai Teka Complex and Its Relationship To
      Easter Island Architectural Prehistory, Social and Linguis-
      tics Institute, University of Hawaii.

Mulloy, W., Liller, W. & Lee, G. (1987). A Survey of Ahu Orientations
      Around the Coast of Easter Island. In preparation.

Niedner, M. B., Jr. & Liller, W. (1987). The IHW Island Network, Sky and
      Telescope, 73, 258.

Routledge, K.S. (1919). The Mystery of Easter Island. London: Hazell,
      Watson, and Viney.Thom, A. (1967). Megalithic Sites in
      Britain. Oxford; Clarendon.

Van Tilburg, Jo Anne (1986). Power and Symbol: The Stylistic Analysis of
      Easter Island Monolithic Sculpture. Ph.D. dissertation,
      University of California, Los Angeles.

Yeomans, D.K., Rahe, J., & Freitag, R.S. (1986). The History of Halley's
      Comet, Journal Royal Astronomical Society Canada, 80, 62.

13   *QUESTIONS*

J. Grindlay: Did the destruction of the monuments precede or follow the contact with the West? It sounded as if the destruction event was nearly simultaneous with contact and, if so, was there a casual connection?

Liller: The Dutch explorer Roggeveen, who first arrived in 1722, said nothing about destruction of monuments, but Captain Cook in 1774 noted that nine statues on three separate platforms "were all fallen down". However, other evidence indicates that the destruction began at least a century before Roggeveen. But it could well be that contact with "civilization" accelerated the process.

NAME INDEX

OBJECT INDEX

SUBJECT INDEX